Sampling Theory and Methods

Sampling Theory and Methods

S. Sampath

CRC Press
Boca Raton London New York Washington, D.C.

Narosa Publishing House
New Delhi Chennai Mumbai Calcutta

S. Sampath
Department of Statistics
Loyola College, Chennai-600 034, India

Library of Congress Cataloging-in-Publication Data:

A catalog record for this book is available from the Library of Congress.

Exclusive distribution in North America only by CRC Press LLC

Direct all inquiries to CRC Press LLC, 2000 N.W. Corporate Blvd., Boca Raton, Florida 33431. E-mail: orders@crcpress.com

International Standard Book Number 0-8493-0980-8
Printed in India.

Dedicated to my parents

Preface

This book is an outcome of nearly two decades of my teaching experience both at the graduate and postgraduate level in Loyola College (Autonomous), Chennai 600 034, during which I came across numerous books and research articles on "Sample Surveys".

I have made an attempt to present the theoretical aspects of "Sample Surveys" in a lucid form for the benefit of both undergraduate and post graduate students of Statistics.

The first chapter of the book introduces to the reader basic concepts of Sampling Theory which are essential to understand the later chapters. Some numerical examples are also presented to help the readers to have clear understanding of the concepts. Simple random sampling design is dealt with in detail in the second chapter. Several solved examples which consider various competing estimators for the population total are also included in the same chapter. The third is devoted to systematic sampling schemes. Various systematic sampling schemes like, linear, circular, balanced, modified systematic sampling and their performances under different superpopulation models are also discussed. In the fourth chapter several unequal probability sampling-estimating strategies are presented. Probability Proportional to Size Sampling With and Without Replacement are considered with appropriate estimators. In addition to them Midzuno sampling scheme and Random group Method are also included. Stratified sampling, allocation problems and related issues are presented with full details in the fifth chapter. Many interesting solved problems are also added. In the sixth and seventh chapters the use of auxiliary information in ratio and regression estimation are discussed. Results related to the properties of ratio and regression estimators under super-population models are also given. Cluster sampling and Multistage sampling are presented in the eighth chapter. The results presented in under two stage sampling are general in nature. In the ninth chapter, non-sampling errors, randomised response techniques and related topics are discussed. Some recent developments in Sample surveys namely, Estimation of distribution functions, Adaptive sampling schemes, Randomised response methods for quantitative data are presented in the tenth chapter.

Many solved theoretical problems are incorporated into almost all the chapters which will help the readers acquire necessary skills to solve problems of theoretical nature on their own.

I am indebted to the authorities of Loyola College for providing me the necessary facilities to successfully complete this work. I also wish to thank Dr.P.Chandrasekar, Department of Statistics, Loyola College, for his help during proof correction. I wish to place on record the excellent work done by the Production Department of Narosa Publishing House in formatting the manuscript

S.Sampath

Contents

Chapter 1

Preliminaries

1.1 Basic Definitions

Definition 1.1 "Finite Population" A finite population is nothing but a set containing finite number of distinguishable elements.

The elements of a finite population will be entities possessing particular characteristics in which a sampler would be interested and they will be referred to as population units. For example, in an agricultural study where one is interested in finding the total yield, a collection of fields or a collection of plots may be defined as population. In a socio-economic study, population units may be defined as a group of individuals, streets or villages.

Definition 1.2 "Population Size" The number of elements in a finite population is called population size. Usually it is denoted by N and it is always a known finite number.

With each unit in a population of size N, a number from 1 through N is assigned. These numbers are called labels of the units and they remain unchanged throughout the study. The values of the population units with respect to the characteristic y under study will be denoted by $Y_1, Y_2, ..., Y_N$. Here Y_i denotes the value of the unit bearing label i with respect to the variable y.

Definition 1.3 "Parameter" Any real valued function of the population values is called parameter.

For example, the population mean $\bar{Y} = \frac{1}{N}\sum_{i=1}^{N} Y_i$, $S^2 = \frac{1}{N-1}\sum_{i=1}^{N}[Y_i - \bar{Y}]^2$ and population range $R = \text{Max}\{X_i\} - \text{Min}\{X_i\}$ are parameters.

Definition 1.4 "Sample" A sample is nothing but a subset of the population S.

Usually it is denoted by s. The number of elements in a sample s is denoted by $n(s)$ and it is referred to as sample size.

Definition 1.5 "Probability Sampling" Choosing a subset of the population according to a probability sampling design is called probability sampling.

Generally a sample is drawn to estimate the parameters whose values are not known.

Definition 1.6 "Statistic" Any real valued function is called statistic, if it depends on $Y_1, Y_2, ..., Y_N$ only through s.

A statistic when used to estimate a parameter is referred to as estimator.

Definition 1.7 "Sampling Design" Let Ω be the collection of all subsets of S and $P(s)$ be a probability distribution defined on Ω. The probability distribution $\{P(s), s \in \Omega\}$ is called sampling design.

A sampling design assigns probability of selecting a subset s as sample. For example, let Ω be the collection of all $\binom{N}{n}$ possible subsets of size n of the population S. The probability distribution

$$P(s) = \begin{cases} \binom{N}{n}^{-1} & \text{if } n(s) = n \\ 0 & \text{otherwise} \end{cases}$$

is a sampling design. This design assigns probabilities $\binom{N}{n}^{-1}$ for all subsets of size n for being selected as sample and zero for all other subsets of S.

It is pertinent to note that the definition of sample as a subset of S does not allow repetition of units in the sample more than once. That is, the sample will always contain distinct units. Alternatively one can also define a sequence whose elements are members of S as a sample, in which case the sample will not necessarily contain distinct units.

Definition 1.8 "Bias" Let $P(s)$ be a sampling design defined on Ω. An estimator $\hat{T}(s)$ is unbiased for the parameter θ with respect to the sampling design $P(s)$ if $E_P[\hat{T}(s)] = \sum_{s \in \Omega} \hat{T}(s) P(s) = \theta$.

The difference $E_P[\hat{T}(s)] - \theta$ is called the bias of $\hat{T}(s)$ in estimating θ with respect to the design $P(s)$. It is to be noted that an estimator which is unbiased with respect to a sampling design $P(s)$ is not necessarily unbiased with respect to some other design $Q(s)$.

Definition 1.9 "Mean Square Error" Mean square error of the estimator $\hat{T}(s)$ in estimating θ with respect to the design $P(s)$ is defined as

$$\begin{aligned} MSE(\hat{T} : P) &= E_P[\hat{T}(s) - \theta]^2 \\ &= \sum_{s \in \Omega} [\hat{T}(s) - \theta]^2 P(s) \end{aligned}$$

If $E_P[\hat{T}(s)] = \theta$ then the mean square error reduces to variance.

Given a parameter θ, one can propose a number of estimators. For example, to estimate the population mean one can use either sample mean or sample median or any other reasonable sample quantity. Hence one requires some criteria to choose an estimator. In sample surveys, we use either the bias or the mean square error or both of them to evaluate the performance of an estimator. Since the bias gives the weighted average of the difference between the estimator and parameter and the mean square error gives weighted squared difference of the estimator and the parameter, it is always better to choose an estimator which has smaller bias (if possible unbiased) and lesser mean square error. The following theorem gives the relationship between the bias and mean square error of an estimator.

Theorem 1.1 Under the sampling design $P(s)$, any statistic $\hat{T}(s)$ satisfies the relation $MSE(P:\hat{T}) = V_P(\hat{T}) + [B_P(\hat{T})]^2$ where $V_P(\hat{T})$ and $B_P(\hat{T})$ are variance and bias of the statistic $\hat{T}(s)$ under the sampling design $P(s)$.

Proof

$$
\begin{aligned}
MSE(\hat{T}:P) &= E_P[\hat{T}(s) - \theta]^2 \\
&= \sum_{s\in\Omega} [\hat{T}(s) - \theta]^2 P(s) \\
&= \sum_{s\in\Omega} [\hat{T}(s) - E_P(\hat{T}(s)) + E_P(\hat{T}(s)) - \theta]^2 P(s) \\
&= \sum_{s\in\Omega} [\hat{T}(s) - E_P(\hat{T}(s))]^2 P(s) + [E_P(\hat{T}(s)) - \theta]^2 \\
&= V_P(\hat{T}) + [B_P(\hat{T})]^2
\end{aligned}
$$

Hence the proof. ■

As mentioned earlier, the performance of an estimator is evaluated on the basis of its bias and mean square error of the estimator. Another way to assess the performance of a sampling design is the use of its *entropy*.

Definition 1.10 "Entropy" Entropy of the sampling design $P(s)$ is defined as,

$$e = -\sum_{s\in\Omega} P(s) \ln P(s)$$

Since the entropy is a measure of information corresponding to the given sampling design, we prefer a sampling design having maximum entropy.

1.2 Estimation of Population Total

In order to introduce the most popular Horvitz-Thompson estimator for the population total we give the following definitions.

Definition 1.11 "Inclusion indicators" Let $s \ni i$ denote the event that the sample s contains the unit i . The random variables

$$I_i(s) = \begin{cases} 1 \text{ if } s \ni i, \ 1 \le i \le N \\ 0 \text{ otherwise} \end{cases}$$

are called inclusion indicators.

Definition 1.12 "Inclusion Probabilities" The first and second order inclusion probabilities corresponding to the sampling design $P(s)$ are defined as

$$\pi_i = \sum_{s \ni i} P(s), \ \pi_{ij} = \sum_{s \ni i,j} P(s)$$

where the sum $\sum_{s \ni i}$ extends over all s containing i and the sum $\sum_{s \ni i,j}$ extends over all s containing both i and j.

Theorem 1.2 For any sampling design $P(s)$, (a) $E_P[I_i(s)] = \pi_i, i = 1, 2, ..., N$ (b) $E_p[I_i(s) I_j(s)] = \pi_{ij}, i, j = 1, 2, ..., N$

Proof (a) Let Ω_1 be the collection of all subsets of S containing the unit with label i and $\Omega_2 = \Omega - \Omega_1$.

$$\begin{aligned} E_P[I_i(s)] &= \sum_{s \in \Omega_1} I_i(s) P(s) + \sum_{s \in \Omega_2} I_i(s) P(s) \\ &= \sum_{s \in \Omega_1} 1 P(s) + \sum_{s \in \Omega_2} 0 P(s) \\ &= \sum_{s \ni i} P(s) \\ &= \pi_i \end{aligned}$$

(b) Let Ω_1 be the collection of all subsets of S containing the units with labels i and j and $\Omega_2 = \Omega - \Omega_1$.

Note that, $I_i(s) I_j(s) = \begin{cases} 1 \text{ if } s \in \Omega_1 \\ 0 \text{ otherwise} \end{cases}$

Therefore

$$\begin{aligned} E_P[I_i(s) I_j(s)] &= \sum_{s \in \Omega_1} I_i(s) I_j(s) P(s) + \sum_{s \in \Omega_2} I_i(s) I_j(s) P(s) \\ &= \sum_{s \in \Omega_1} P(s) \\ &= \sum_{s \ni i,j} P(s) \\ &= \pi_{ij} \end{aligned}$$

Hence the proof. ■

Theorem 1.3 For any sampling design $P(s)$, $E_P[n(s)] = \sum_{i=1}^{N} \pi_i$

Proof For any sampling design, we know that,

$$n(s) = \sum_{i=1}^{N} I_i(s)$$

Taking expectation on both sides, we get

$$E_P[n(s)] = \sum_{i=1}^{N} E_P[I_i(s)]$$

$$= \sum_{i=1}^{N} \pi_i$$

Hence the proof. ■

Theorem 1.4 (a) For $i = 1, 2, \ldots, N$, $V_P[I_i(s)] = \pi_i(1-\pi_i)$,
(b) For $i, j = 1, 2, \ldots, N$, $\text{cov}_P[I_i(s), I_j(s)] = \pi_{ij} - \pi_i\pi_j$
Proof of this theorem is straight forward and hence left as an exercise.

Theorem 1.5 Under any sampling design, satisfying $P[n(s) = n] = 1$ for all s,

(a) $n = \sum_{i=1}^{N} \pi_i$ (b) $\sum_{j \neq i}^{N} [\pi_i\pi_j - \pi_{ij}] = \pi_i(1-\pi_i)$

Proof (a) Proof of this part follows from Theorem 10.3

(b) Since for every s, $P[n(s) = n] = 1$, we have $\sum_{j \neq i}^{N} I_j(s) = n - I_i(s)$

Hence by Theorem 1.4, we write

$$\pi_i(1-\pi_i) = V_P[I_i(s)]$$

$$= \text{cov}_P[I_i(s), I_i(s)]$$

$$= \text{cov}_P[I_i(s), n - \sum_{j \neq i}^{N} I_j(s)]$$

$$= -\sum_{j \neq i}^{N} \text{cov}_P[I_i(s), I_j(s)]$$

$$= \sum_{j \neq i}^{N} [\pi_i\pi_j - \pi_{ij}]$$

Hence the proof. ■

Using the first order inclusion probabilities, Horvitz and Thompson (1952) constructed an unbiased estimator for the population total. Their estimator for the population total is $\hat{Y}_{HT} = \sum_{i \in s} \frac{Y_i}{\pi_i}$. The following theorem proves that the above estimator is unbiased for the population total and also it gives the variance.

Theorem 1.6 The Horvitz-Thompson estimator $\hat{Y}_{HT} = \sum_{i \in s} \frac{Y_i}{\pi_i}$ is unbiased for the population total and its variance is

$$\sum_{i=1}^{N} Y_i^2 \left[\frac{1-\pi_i}{\pi_i}\right] + 2\sum_{i=1}^{N}\sum_{\substack{j=1 \\ i<j}}^{N} Y_i Y_j \left[\frac{\pi_{ij} - \pi_i \pi_j}{\pi_i \pi_j}\right]$$

Proof The estimator $\hat{Y}_{HT}$ can be written as

$$\hat{Y}_{HT} = \sum_{i=1}^{N} \frac{Y_i}{\pi_i} I_i(s)$$

Taking expectations on both sides we get

$$E_P[\hat{Y}_{HT}] = \sum_{i=1}^{N} \frac{Y_i}{\pi_i} \pi_i = Y$$

Therefore $\hat{Y}_{HT}$ is unbiased for the population total.
Consider the difference

$$\hat{Y}_{HT} - Y = \sum_{i=1}^{N} \frac{Y_i}{\pi_i} I_i(s) - \sum_{i=1}^{N} Y_i$$

$$= \sum_{i=1}^{N} \frac{Y_i}{\pi_i} [I_i(s) - \pi_i]$$

Squaring both the sides and taking expectations we get

$$E_P[\hat{Y}_{HT} - Y]^2 = \sum_{i=1}^{N} \left[\frac{Y_i}{\pi_i}\right]^2 E_P[I_i(s) - \pi_i]^2$$

$$+ 2\sum_{i=1}^{N}\sum_{\substack{j=1 \\ i<j}}^{N} \frac{Y_i}{\pi_j}\frac{Y_j}{\pi_j} E_P[I_i(s) - \pi_i][I_j(s) - \pi_j]$$

$$= \sum_{i=1}^{N} \left[\frac{Y_i}{\pi_i}\right]^2 V_P[I_i(s)] + 2\sum_{i=1}^{N}\sum_{\substack{j=1 \\ i<j}}^{N} \frac{Y_i}{\pi_j}\frac{Y_j}{\pi_j} \operatorname{cov}_P(I_i(s), I_j(s))$$

$$= \sum_{i=1}^{N} Y_i^2 \left[\frac{1-\pi_i}{\pi_i}\right] + 2\sum_{i=1}^{N}\sum_{\substack{j=1 \\ i<j}}^{N} Y_i Y_j \left[\frac{\pi_{ij} - \pi_i \pi_j}{\pi_i \pi_j}\right]$$

Hence the proof. ■

Remark The variance of Horvitz-Thompson estimator can also be expressed in the following form

$$\sum_{i=1}^{N}\sum_{\substack{j=1\\i<j}}^{N}\left[\frac{Y_i}{\pi_i}-\frac{Y_j}{\pi_j}\right]^2[\pi_i\pi_j-\pi_{ij}]$$

Proof for Remark From the previous theorem, we have

$$V_p[\hat{Y}_{HT}]=\sum_{i=1}^{N}Y_i^2\left[\frac{1-\pi_i}{\pi_i}\right]+2\sum_{i=1}^{N}\sum_{\substack{j=1\\i<j}}^{N}Y_iY_j\left[\frac{\pi_{ij}-\pi_i\pi_j}{\pi_i\pi_j}\right]$$

$$=\sum_{i=1}^{N}\left[\frac{Y_i^2}{\pi_i^2}\right]\sum_{\substack{j=1\\j\neq i}}^{N}[\pi_i\pi_j-\pi_{ij}]+\sum_{i=1}^{N}\sum_{\substack{j=1\\j\neq i}}^{N}Y_iY_j\left[\frac{\pi_{ij}-\pi_i\pi_j}{\pi_i\pi_j}\right]$$

$$=\frac{1}{2}\sum_{i=1}^{N}\sum_{\substack{j=1\\j\neq i}}^{N}\left(\frac{Y_i^2}{\pi_i^2}+\frac{Y_j^2}{\pi_j^2}\right)[\pi_i\pi_j-\pi_{ij}]-\sum_{i=1}^{N}\sum_{\substack{j=1\\j\neq i}}^{N}\frac{Y_i}{\pi_i}\frac{Y_j}{\pi_j}[\pi_i\pi_j-\pi_{ij}]$$

$$=\frac{1}{2}\sum_{i=1}^{N}\sum_{\substack{j=1\\j\neq i}}^{N}\left(\frac{Y_i}{\pi_i}-\frac{Y_j}{\pi_j}\right)^2[\pi_i\pi_j-\pi_{ij}]$$

$$=\sum_{i=1}^{N}\sum_{\substack{j=1\\i<j}}^{N}\left[\frac{Y_i}{\pi_i}-\frac{Y_j}{\pi_j}\right]^2[\pi_i\pi_j-\pi_{ij}]$$

Hence the proof. ■

The above form of the variance of the Horvitz-Thompson estimator is known as Yates-Grundy form. It helps us to get an unbiased estimator of the variance of Horvitz-Thompson Estimator very easily. Consider any design yielding positive second order inclusion probabilities for all pairs of units in the population. For any such design, an unbiased estimator of the variance given above is

$$\sum_{i\in s}\sum_{\substack{j\in s\\i<j}}\left[\frac{Y_i}{\pi_i}-\frac{Y_j}{\pi_j}\right]^2\left(\frac{\pi_i\pi_j-\pi_{ij}}{\pi_{ij}}\right)$$

The Horvitz-Thompson estimator, its variance and also the estimated variance can be used for any sampling design yielding positive first order inclusion probabilities for all the population units and positive second order inclusion probabilities for all pairs of units. It is pertinent to note that the estimated variance is likely to take negative values. A set of sufficient conditions for the non-negativity of the estimated variance expression given above are, for every pair of units i and j, $\pi_i\pi_j-\pi_{ij}\geq 0$ ($i, j=1,2,\ldots,N$).

Note For estimating population total, apart from Horvitz-Thompson estimator several other estimators are also available in literature and some of them are presented in later chapters at appropriate places.

1.3 Problems and Solutions

Problem 1.1 Show that a necessary and sufficient condition for unbiased estimation of a finite population total is that the first order inclusion probabilities must be positive for all units in the population.

Solution When the first order inclusion probabilities are positive for all units, one can use Horvitz-Thompson estimator as an unbiased estimator for the population total.

When the first order inclusion probability is zero for a particular unit say, the unit with label i expected value of any statistic under the given sampling design will be free from Y_i, its value with respect to the variable under study. Hence the first order inclusion probability must be positive for all units in the population. ■

Problem 1.2 Derive $\text{cov}_P(\hat{Y}, \hat{X})$ where $\hat{Y}$ and $\hat{X}$ are Horvitz-Thompson estimators of Y and X, totals of the population units with respect to variables y and x respectively.

Solution For $i = 1, 2, \ldots, N$, let $Z_i = X_i + Y_i$.

Note that $\hat{Z} = \hat{X} + \hat{Y}$

Therefore $V_P[\hat{Z}] = V_P[\hat{X}] + V_P[\hat{Y}] + 2\,\text{cov}_P[\hat{X}, \hat{Y}]$ (1.1)

By remark given under Theorem 1.6, we have

$$V_P[\hat{Z}] = \sum_{i=1}^{N}\sum_{\substack{j=1\\i<j}}^{N}\left[\frac{Z_i}{\pi_i} - \frac{Z_j}{\pi_j}\right]^2 [\pi_i\pi_j - \pi_{ij}]$$

$$= \sum_{i=1}^{N}\sum_{\substack{j=1\\i<j}}^{N}\left[\frac{X_i}{\pi_i} - \frac{X_j}{\pi_j}\right]^2 [\pi_i\pi_j - \pi_{ij}] + \sum_{i=1}^{N}\sum_{\substack{j=1\\i<j}}^{N}\left[\frac{Y_i}{\pi_i} - \frac{Y_j}{\pi_j}\right]^2 [\pi_i\pi_j - \pi_{ij}]$$

$$+ 2\sum_{i=1}^{N}\sum_{\substack{j=1\\i<j}}^{N}\left[\frac{Y_i}{\pi_i} - \frac{Y_j}{\pi_j}\right]\left[\frac{X_i}{\pi_i} - \frac{X_j}{\pi_j}\right][\pi_i\pi_j - \pi_{ij}] \quad (1.2)$$

Comparing (1.1) and (1.2) we get

$$\text{cov}_P(\hat{X}, \hat{Y}) = \sum_{i=1}^{N}\sum_{\substack{j=1\\i<j}}^{N}\left[\frac{Y_i}{\pi_i} - \frac{Y_j}{\pi_j}\right]\left[\frac{X_i}{\pi_i} - \frac{X_j}{\pi_j}\right][\pi_i\pi_j - \pi_{ij}]$$

Hence the solution. ■

Exercises

1.1 Let S be a finite population containing 5 units. Calculate all first and second order inclusion probabilities under the sampling design

$$P(s)=\begin{cases}0.2 & \text{for } s=\{2,3,4\}, s=\{2,5\}\\ 0.3 & \text{for } s=\{1,3,5\}, s=\{1,4\}\\ 0 & \text{otherwise}\end{cases}$$

1.2 From a population containing five units with values 4,7,11,17 and 23, four units are drawn with the help of the design

$$P(s)=\begin{cases}0.20 & \text{if } n(s)=5\\ 0 & \text{otherwise}\end{cases}$$

Compare the bias and mean square error of sample mean and median in estimating the population mean.

1.3 List all possible values of $n(s)$ under the sampling design given in Problem 1.1 and verify the relation $E_P[n(s)]=\sum_{i=1}^{N}\pi_i$

1.4 Check the validity of the statement "Under any sampling design the sum of first order inclusion probabilities is always equal to sample size".

1.5 Check the validity of the statement "Horvitz-Thompson estimator can be used under any sampling design to obtain an unbiased estimate for the population total".

Chapter 2

Equal Probability Sampling

2.1 Simple Random Sampling

This is one of the simplest and oldest methods of drawing a sample of size n from a population containing N units. Let Ω be the collection of all 2^N subsets of S. The probability sampling design

$$P(s)=\begin{cases}\binom{N}{n}^{-1} & \text{if } n(s)=n \\ 0 & \text{otherwise}\end{cases}$$

is known as simple random sampling design.

In the above design each one of the $\binom{N}{n}$ possible sets of size n is given equal probability for being selected as sample. The above design can be implemented by following the unit drawing procedure described below:

Choose n random numbers from 1 through N without replacement. Units corresponding to the numbers selected as above will constitute the sample.

Now we shall prove this sampling mechanism implements the sampling design defined above.

Consider an arbitrary subset s of the population S whose members are i_1, i_2, i_3, ..., i_n. The probability of selecting the units $i_1, i_2, i_3, \ldots, i_n$ in the order $i_1 \to i_2 \to i_3 \to \ldots \to i_n$ is

$$\frac{1}{N}\frac{1}{N-1}\frac{1}{N-2}\cdots\frac{1}{N-(n-1)}.$$

Since the number of ways in which these n units can be realized is $n!$, the probability of obtaining the set s as sample is

$$n!\frac{1}{N}\frac{1}{N-1}\frac{1}{N-2}\cdots\frac{1}{N-(n-1)}$$

which reduces on simplification to $\binom{N}{n}^{-1}$.

Therefore we infer that the sampling mechanism described above will implement the simple random sampling design.

2.2 Estimation of Total

The following theorem gives the first and second order inclusion probabilities under simple random sampling.

Theorem 2.1 Under simple random sampling, (a) For $i = 1, 2, ..., N$, $\pi_i = \frac{n}{N}$

(b) For $i, j = 1, 2, ..., N$, $\pi_{ij} = \frac{n(n-1)}{N(N-1)}$

Proof By definition

$$\pi_i = \sum_{s \ni i} P(s) = \sum_{s \ni i} \binom{N}{n}^{-1}$$

Since there are $\binom{N-1}{n-1}$ subsets with i as an element, the above sum reduces to $\binom{N-1}{n-1}\binom{N}{n}^{-1}$ which is equal to $\frac{n}{N}$

Again by definition, we have

$$\pi_{ij} = \sum_{s \ni i,j} P(s) = \sum_{s \ni i,j} \binom{N}{n}^{-1}$$

Since there are $\binom{N-2}{n-2}$ subsets with i and j as elements, we get

$$\pi_{ij} = \binom{N-2}{n-2}\binom{N}{n}^{-1} = \frac{n(n-1)}{N(N-1)}$$

Hence the proof. ■

The following theorem gives an unbiased estimator for the population total and also its variance.

Theorem 2.2 Under simple random sampling, $\hat{Y}_{srs} = \frac{N}{n}\sum_{i \in s} Y_i$ is unbiased for the population total and its variance is $V[\hat{Y}_{srs}] = \frac{N^2(N-n)}{Nn} S_y^2$ where $S_y^2 = \frac{1}{N-1}\sum_{i=1}^{N}[Y_i - \bar{Y}]^2$

Proof We have seen in Chapter 1, for any sampling design

$$\hat{Y}_{HT} = \sum_{i \in s} \frac{Y_i}{\pi_i} \tag{2.1}$$

is unbiased for the population total with variance

$$\sum_{i=1}^{N}\sum_{\substack{j=1\\i<j}}^{N}\left[\frac{Y_i}{\pi_i}-\frac{Y_j}{\pi_j}\right]^2(\pi_i\pi_j-\pi_{ij}) \tag{2.2}$$

By Theorem 2.1, we have $\pi_i = \dfrac{n}{N}$ and $\pi_{ij} = \dfrac{n(n-1)}{N(N-1)}$

Substituting these values in (2.1) we notice that

$$\hat{Y}_{HT} = \frac{N}{n}\sum_{i\in s} Y_i \tag{2.3}$$

is unbiased for the population total.
Note that

$$\pi_i\pi_j - \pi_{ij} = \frac{n^2}{N^2} - \frac{n(n-1)}{N(N-1)} = \frac{n}{N}\frac{N-n}{N(N-1)} \tag{2.4}$$

Therefore by (2.2) $V(\hat{Y}_{HT}) = \displaystyle\sum_{i=1}^{N}\sum_{\substack{j=1\\i<j}}^{N}\left[\frac{N^2}{n^2}\right]^2 (Y_i - Y_j)^2 \frac{n(N-n)}{N(N-1)}$

$$= \frac{N-n}{n(N-1)}\sum_{i=1}^{N}\sum_{\substack{j=1\\i<j}}^{N}(Y_i - Y_j)^2 \tag{2.5}$$

We know that $\displaystyle\sum_{i=1}^{N}\sum_{j=1}^{N} a_{ij} = \sum_{i=1}^{N} a_{ii} + 2\sum_{i=1}^{N}\sum_{\substack{j=1\\i<j}}^{N} a_{ij}$, if $a_{ij} = a_{ji}$.

Using the above identity in the right hand side of (2.5), we get

$$V(\hat{Y}_{HT}) = \frac{N-n}{2n(N-1)}\left\{\sum_{i=1}^{N}\sum_{j=1}^{N}(Y_i - Y_j)^2 - \sum_{i=1}^{N}(Y_i - Y_i)^2\right\}$$

$$= \frac{N-n}{2n(N-1)}\left\{2\sum_{i=1}^{N}\sum_{j=1}^{N} Y_i^2 - 2\sum_{i=1}^{N}\sum_{j=1}^{N} Y_i Y_j\right\}$$

$$= \frac{N-n}{n(N-1)}\left\{\sum_{i=1}^{N} NY_i^2 - N^2\overline{Y}^2\right\}$$

$$= \frac{N(N-n)}{n(N-1)}\sum_{i=1}^{N}(Y_i - \overline{Y})^2 = \frac{N^2(N-n)}{Nn} S_y^2 \tag{2.6}$$

Hence the proof. ■

The following theorem gives an unbiased estimator for the variance obtained in Theorem 2.1.

Theorem 2.3 An unbiased estimator of $V(\hat{Y}_{srs})$ is $v(\hat{Y}_{srs}) = \dfrac{N^2(N-n)}{Nn} s_y^2$ where s_y^2 is the sample analogue of S_y^2.

Proof Since $V(\hat{Y}_{srs}) = E(\hat{Y}_{srs}^2) - Y^2$,

we have $E(\hat{Y}_{srs}^2) = V(\hat{Y}_{srs}) + Y^2 = \dfrac{N^2(N-n)}{Nn} S_y^2 + Y^2$ (2.7)

The sample analogue of S_y^2 is $s_y^2 = \dfrac{1}{n-1}\sum_{i\in s}[Y_i - \hat{\bar{Y}}]^2$ where $\hat{\bar{Y}} = \dfrac{1}{n}\sum_{i\in s} Y_i$

$$= \frac{1}{n-1}\{\sum_{i\in s} Y_i^2 - n\hat{\bar{Y}}^2\}$$

$$= \frac{1}{n-1}\{\sum_{i\in s} Y_i^2 - n\left[\frac{\hat{Y}_{srs}^2}{N^2}\right]\}$$

Taking expectations on both sides we get

$$E(s_y^2) = \frac{1}{n-1}\left\{E\left[\sum_{i\in s} Y_i^2\right] - nE[\hat{\bar{Y}}^2]\right\}$$

$$= \frac{1}{n-1}\left[\frac{n}{N}\sum_{i=1}^{N} Y_i^2 - n\left\{\frac{N-n}{Nn} S_y^2 + \bar{Y}^2\right\}\right] \text{(using (2.7))}$$

$$= \frac{n}{n-1}\left[\frac{1}{N}\sum_{i=1}^{N} Y_i^2 - \left\{\frac{N-n}{Nn} S_y^2 + \bar{Y}^2\right\}\right]$$

$$= \frac{n}{n-1}\left[\frac{N-1}{N} S_y^2 - \frac{N-n}{Nn} S_y^2\right]$$

$$= \frac{n}{n-1}\left[\frac{n-1}{n} S_y^2\right] = S_y^2 \qquad (2.8)$$

This implies $E\left[\dfrac{N^2(N-n)}{Nn} s_y^2\right] = \left[\dfrac{N^2(N-n)}{Nn}\right] S_y^2$

Hence the proof. ■

Theorem 2.4 Let (X_i, Y_i) be the values with respect to the two variables x and y associated with the unit having label i, $i = 1, 2, \ldots, N$. If $\hat{X} = \dfrac{N}{n}\sum_{i\in s} X_i$,

$\hat{Y} = \frac{N}{n}\sum_{i\in s} Y_i$ and $S_{xy} = \frac{1}{N-1}\sum_{i=1}^{N}(X_i - \overline{X})(Y_i - \overline{Y})$, then under simple random sampling, $\text{cov}(\hat{X},\hat{Y}) = \left[\frac{N^2(N-n)}{Nn}\right]S_{xy}$.

Proof

$$\begin{aligned}\text{cov}(\hat{X},\hat{Y}) &= E[\hat{X} - E(\hat{X})][\hat{Y} - E(\hat{Y})] \\ &= E[\hat{X}\,\hat{Y}] - E(\hat{X})E(\hat{Y}) \\ &= E\left[\left(\frac{N}{n}\right)^2 \sum_{i\in s} X_i \sum_{i\in s} Y_i\right] - XY \\ &= \left(\frac{N}{n}\right)^2 E\left[\sum_{i\in s} Y_i X_i + \sum_{\substack{i,j\in s \\ i\neq j}} Y_i X_j\right] - XY \\ &= \left(\frac{N}{n}\right)^2 \left\{\frac{n}{N}\sum_{i=l}^{N} Y_i X_i + \frac{n(n-1)}{N(N-1)}\sum_{i\neq j}^{N} Y_i X_j\right\} - XY \qquad (2.9)\end{aligned}$$

It is well known that $\sum_{i=1}^{N} Y_i \sum_{j=1}^{N} X_j = \sum_{i=1}^{N} Y_i X_i + \sum_{i=1}^{N}\sum_{i\neq j}^{N} Y_i X_j$

Hence

$$N^2\overline{Y}\,\overline{X} - \sum_{i=1}^{N} Y_i X_i = \sum_{i=1}^{N}\sum_{i\neq j}^{N} Y_i X_j \qquad (2.10)$$

Substituting (2.10) in (2.9) we get

$$\begin{aligned}\text{cov}(\hat{X},\hat{Y}) &= \left[\frac{N}{n}\right]^2 \left[\frac{n}{N}\right]\sum_{i=1}^{N} X_i Y_i + \left[\frac{n(n-1)}{N(N-1)}\right]\left[N^2\overline{Y}\,\overline{X} - \sum_{i=1}^{N} Y_i X_i\right] - YX \\ &= \left[\frac{N}{n}\right]^2 \left[\frac{n}{N}\right]\left[1 - \frac{n(n-1)}{N(N-1)}\right]\sum_{i=1}^{N} X_i Y_i + \left[\frac{n(n-1)}{N(N-1)}\right]\left[N^2\overline{Y}\,\overline{X}\right] - N^2\overline{Y}\,\overline{X} \\ &= \frac{N^2(N-n)}{Nn}\frac{1}{N-1}\sum_{i=1}^{N}(X_i - \overline{X})(Y_i - \overline{Y})\end{aligned}$$

Hence the proof. ■

Theorem 2.5 Under simple random sampling $s_{xy} = \frac{1}{n-1}\sum_{i\in s}(X_i - \hat{\overline{X}})(Y_i - \hat{\overline{Y}})$

is unbiased for $S_{xy} = \frac{1}{N-1}\sum_{i=1}^{N}(X_i - \overline{X})(Y_i - \overline{Y})$ where $\hat{\overline{X}} = \frac{1}{n}\sum_{i\in s} X_i$ and $\hat{\overline{Y}} = \frac{1}{n}\sum_{i\in s} Y_i$.

Proof $s_{xy} = \frac{1}{n-1}\left[\sum_{i\in s} X_i Y_i - n\hat{\overline{X}}\hat{\overline{Y}}\right]$

$$= \frac{1}{n-1}\left[\sum_{i\in s} X_i Y_i - \frac{n}{n^2}\sum_{i\in s} X_i \sum_{i\in s} Y_i\right]$$

$$= \frac{1}{n-1}\left[\sum_{i\in s} X_i Y_i - \frac{1}{n}\left\{\sum_{i\in s} Y_i X_i + \sum_{i\in s}\sum_{\substack{i\neq j\\ j\in s}} Y_i X_i\right\}\right]$$

$$= \frac{1}{n-1}\left\{\frac{n-1}{n}\sum_{i\in s} Y_i X_i - \frac{1}{n}\sum_{i\in s}\sum_{\substack{j\in s\\ i\neq j}} Y_i X_j\right\}$$

Taking expectations on both sides, we get

$$E[s_{xy}] = \frac{1}{n-1}\left[\frac{n-1}{n}\frac{n}{N}\sum_{i=1}^{N} Y_i X_i - \frac{n(n-1)}{nN(N-1)}\sum_{i=1}^{N}\sum_{\substack{j=1\\ j\neq i}}^{N} Y_i X_j\right]$$

$$= \frac{1}{N}\sum_{i=1}^{N} Y_i X_i - \frac{1}{N(N-1)}\sum_{i=1}^{N}\sum_{\substack{j=1\\ j\neq i}}^{N} Y_i X_j$$

$$= \frac{1}{N}\sum_{i=1}^{N} Y_i X_i - \frac{1}{N(N-1)}\left[N^2\overline{Y}\,\overline{X} - \sum_{i=1}^{N} Y_i X_i\right]$$

$$= \left[\frac{1}{N} + \frac{1}{N(N-1)}\right]\sum_{i=1}^{N} Y_i X_i - \left[\frac{N}{N-1}\right]\overline{Y}\,\overline{X}$$

$$= \frac{1}{N-1}\left[\sum_{i=1}^{N} Y_i X_i - N\overline{Y}\,\overline{X}\right]$$ Hence the proof. ■

Remark 2.1 If $\hat{\overline{Y}} = \frac{\hat{Y}_{srs}}{N}$ then under simple random sampling $\hat{\overline{Y}}$ is unbiased for the population mean and its variance is $\frac{N-n}{Nn}S_y^2$.

This remark follows from Theorem 2.2.

2.3 Problems and Solutions

Problem 2.1 After the decision to take a simple random sample had been made, it was realised that Y_1 the value of unit with label 1 would be unusually low and Y_N the value of the unit with label N would be unusually high. In such situations, it is decided to use the estimator

$$\hat{\bar{Y}}^* = \begin{cases} \hat{\bar{Y}} + C \text{ if the sample contains } Y_N \text{ but not } Y_1 \\ \hat{\bar{Y}} - C \text{ if the sample contains } Y_1 \text{ but not } Y_N \\ \hat{\bar{Y}} \text{ for all other samples} \end{cases}$$

where the constant C is positive and predetermined. Show that the estimator $\hat{\bar{Y}}^*$ is unbiased and its variance is

$$V(\hat{\bar{Y}}^*) = \frac{N-n}{N}\left[\frac{S_y^2}{n} - \frac{2C}{N-1}(Y_N - Y_1 - nC)\right]$$

Also prove that $V(\hat{\bar{Y}}^*) < V(\hat{\bar{Y}})$ if $0 < C < \dfrac{Y_N - Y_1}{n}$ (Sarndal (1971)).

Solution Let $\Omega_n = \{s / n(s) = n\}$ Partition Ω_n into three disjoint subclasses as
$\Omega_1 = \{s / n(s) = n, s \text{ contains } 1 \text{ but not } N\}$,
$\Omega_2 = \{s / n(s) = n, s \text{ contains } N \text{ but not } 1\}$
and $\Omega_3 = \Omega_n - \Omega_1 - \Omega_2$

It is to be noted that the number of subsets in Ω_1, Ω_2 and Ω_3 are respectively $\binom{N-2}{n-1}, \binom{N-2}{n-1}$ and $\binom{N}{n} - 2\binom{N-2}{n-1}$.

Under simple random sampling

$$E(\hat{\bar{Y}}^*) = \sum_{s \in \Omega_n} \hat{\bar{Y}}^* \binom{N}{n}^{-1}$$

$$= \binom{N}{n}^{-1}\left\{\sum_{s\in\Omega_1}\left[\hat{\bar{Y}} + C\right] + \sum_{s\in\Omega_2}\left[\hat{\bar{Y}} - C\right] + \sum_{s\in\Omega_3}\hat{\bar{Y}}\right\}$$

$$= \binom{N}{n}^{-1}\left\{\sum_{s\in\Omega_n}\hat{\bar{Y}} + C\binom{N-2}{n-1} - C\binom{N-2}{n-1}\right\}$$

$$= \binom{N}{n}^{-1}\sum_{s\in\Omega_n}\hat{\bar{Y}} = \bar{Y} \text{ (refer the remark 2.1)}$$

Therefore the estimator $\hat{\bar{Y}}^*$ is unbiased for the population mean. The variance of the estimator $\hat{\bar{Y}}^*$ is $V(\hat{\bar{Y}}^*) = \sum_{s\in\Omega_n}\left[\hat{\bar{Y}}^* - \bar{Y}\right]^2\binom{N}{n}^{-1}$ (by definition)

$$=\binom{N}{n}^{-1}\left(\sum_{s\in\Omega_1}\left[\hat{\bar{Y}}+C-\bar{Y}\right]^2+\sum_{s\in\Omega_2}\left[\hat{\bar{Y}}-C-\bar{Y}\right]^2+\sum_{s\in\Omega_3}\left[\hat{\bar{Y}}-\bar{Y}\right]^2\right)$$

$$=\binom{N}{n}^{-1}\Bigg[\sum_{s\in\Omega_1}\left[\hat{\bar{Y}}-\bar{Y}\right]^2+\sum_{s\in\Omega_2}\left[\hat{\bar{Y}}-\bar{Y}\right]^2+\sum_{s\in\Omega_3}\left[\hat{\bar{Y}}-\bar{Y}\right]^2$$

$$+C^2\left\{\binom{N-2}{n-1}+\binom{N-2}{n-1}\right\}-2C\left\{\sum_{s\in\Omega_2}[\hat{\bar{Y}}-\bar{Y}]-\sum_{s\in\Omega_1}[\hat{\bar{Y}}-\bar{Y}]\right\}\Bigg]$$

Note that $\dfrac{\binom{N-2}{n-1}+\binom{N-2}{n-1}}{\binom{N}{n}}=\dfrac{2n(N-n)}{N(N-1)}$. Further it may be noted that all the $\binom{N-2}{n-1}$ members of Ω_1 contain the unit with label 1, $\binom{N-3}{n-2}$ of them contain the units with labels j $(j=2,3,\ldots,N-1)$ and none of them contain the unit with label N. Therefore

$$\sum_{s\in\Omega_1}[\hat{\bar{Y}}-\bar{Y}]=\sum_{s\in\Omega_1}\hat{\bar{Y}}-\sum_{s\in\Omega_1}\bar{Y}$$

$$=\frac{1}{n}\left[\binom{N-2}{n-1}Y_1+\binom{N-3}{n-2}\sum_{j=2}^{N-1}Y_j-\binom{N-2}{n-1}\bar{Y}\right]$$

$$=\frac{1}{n}\binom{N-2}{n-1}\left\{Y_1+\frac{n-1}{N-2}\sum_{j=2}^{N-1}Y_j\right\}-\binom{N-2}{n-1}\bar{Y}$$

Proceeding in the same way, we get

$$\sum_{s\in\Omega_2}[\hat{\bar{Y}}-\bar{Y}]=\frac{1}{n}\binom{N-2}{n-1}\left\{Y_N+\frac{n-1}{N-2}\sum_{j=2}^{N-1}Y_j\right\}-\binom{N-2}{n-1}\bar{Y} \tag{2.12}$$

It can be seen that

$$\frac{1}{n}\frac{\binom{N-2}{n-1}}{\binom{N}{n}}=\frac{N-n}{N(N-1)} \tag{2.13}$$

Using (2.10)-(2.13) in (2.9) we get

$$V(\hat{\bar{Y}}^*) = \frac{N-n}{N}\left[\frac{S_y^2}{n} - \frac{2C}{N-1}(Y_N - Y_1 - nC)\right] \tag{2.14}$$

which is the required result.

$$V(\hat{\bar{Y}}) = \frac{N-n}{N}\left[\frac{S_y^2}{n}\right] \tag{2.15}$$

Therefore

$$V(\hat{\bar{Y}}^*) < V(\hat{\bar{Y}}) \Rightarrow \left[\frac{2C}{N-1}\right][Y_N - Y_1 - nC] > 0 \quad \text{(comparing (2.14) and (2.15))}$$

$$\Rightarrow Y_N - Y_1 > nC \quad \text{(when } C \text{ is positive)}$$

$$\Rightarrow 0 < C < \left[\frac{Y_N - Y_1}{n}\right]$$

Hence the solution. ■

Problem 2.2 Given the information in problem 2.1, an alternative plan is to include both Y_1 and Y_8 in every sample, drawing a sample of size 2 from the units with labels 2, 3, ..., 7, when N=8 and n=4. Let $\hat{\bar{Y}}_2$ be the mean of those 2 units selected. Show that the estimator $\hat{\bar{Y}}' = \frac{Y_1 + 6\hat{\bar{Y}}_2 + Y_8}{8}$ is unbiased for the population mean with variance $9\frac{V(\hat{\bar{Y}}_2)}{16}$.

Solution $\hat{\bar{Y}}' = \frac{Y_1 + 6\hat{\bar{Y}}_2 + Y_8}{8} = \frac{Y_1 + Y_8}{8} + \left[\frac{6}{8}\right]\frac{1}{2}\sum_{i \in s} Y_i$

Taking expectations on both the sides we get

$$E\left[\hat{\bar{Y}}'\right] = \left[\frac{Y_1 + Y_8}{8}\right] + \left[\frac{3}{8}\right] E\left[\sum_{i=2}^{7} I_i Y_i\right] \tag{2.16}$$

where $I_i = 1$ if $i \in s$

$= 0$ otherwise

Since $E[I_i] = \frac{2}{6}$, we get from (2.16)

$$E\left[\hat{\bar{Y}}'\right] = \left[\frac{Y_1 + Y_8}{8}\right] + \left[\frac{3}{8}\right]\left[\frac{2}{6}\right]\left[\sum_{i=2}^{7} Y_i\right] = \bar{Y}$$

$$V[\hat{\bar{Y}}'] = V\left\{\frac{Y_1 + 6\hat{\bar{Y}}_2 + Y_8}{8}\right\} = \frac{9}{16} V[\hat{\bar{Y}}_2]$$

Hence the solution. ■

Problem 2.3 Show that when $N = 3, n = 2$ in simple random sampling, the estimator

$$\hat{\bar{Y}}^* = \begin{cases} \frac{1}{2}Y_1 + \frac{1}{2}Y_2 \text{ if } s = \{1,2\} \\ \frac{1}{2}Y_1 + \frac{2}{3}Y_3 \text{ if } s = \{1,3\} \\ \frac{1}{2}Y_2 + \frac{1}{3}Y_3 \text{ if } s = \{2,3\} \end{cases}$$

is unbiased for the population mean and

$$V(\hat{\bar{Y}}^*) > V(\hat{\bar{Y}}) \quad \text{if } Y_3[3Y_2 - 3Y_1 - Y_3] > 0$$

Solution By definition

$$E[\hat{\bar{Y}}^*] = \sum_{s=s_1,s_2,s_3} [\hat{\bar{Y}}^*] \frac{1}{\binom{3}{2}} = \left(\frac{1}{3}\right)\left\{\frac{1}{2}Y_1 + \frac{1}{2}Y_2 + \frac{1}{2}Y_1 + \frac{2}{3}Y_3 + \frac{1}{2}Y_2 + \frac{1}{3}Y_3\right\}$$

$$= \left(\frac{1}{3}\right)\{Y_1 + Y_2 + Y_3\} = \bar{Y}$$

Hence $\hat{\bar{Y}}^*$ is unbiased for the population mean.

$$V[\hat{\bar{Y}}^*] = \frac{1}{3}\left\{\left[\frac{1}{2}Y_1 + \frac{1}{2}Y_2\right]^2 + + \left[\frac{1}{2}Y_1 + \frac{2}{3}Y_3\right]^2 + \left[\frac{1}{2}Y_2 + \frac{1}{3}Y_3\right]^2\right\}$$

$$- \left[\frac{Y_1 + Y_2 + Y_3}{3}\right]^2$$

$$= \frac{1}{18}Y_1^2 + \frac{1}{18}Y_2^2 + \frac{2}{27}Y_3^2 - \frac{1}{18}Y_1Y_2 - \frac{1}{9}Y_2Y_3$$

We know that under simple random sampling,

$$V[\hat{\bar{Y}}] = \frac{3-2}{(3)(2)}\frac{1}{3-1}\sum_{i=1}^{3}[Y_i - \bar{Y}]^2 \qquad \text{(refer remark 2.1)}$$

$$= \frac{1}{18}\left[\, Y_1^2 + Y_2^2 + Y_3^2 - Y_1Y_2 - Y_1Y_3 - Y_2Y_3 \,\right]$$

Therefore

$$V[\hat{\bar{Y}}] - V[\hat{\bar{Y}}^*] = -\left[\frac{1}{54}\right]Y_3^2 + \left[\frac{3}{54}\right]Y_2Y_3 - \left[\frac{3}{54}\right]Y_1Y_3$$

Using the above difference we get,

$$V[\hat{\bar{Y}}] - V[\hat{\bar{Y}}^*] > 0$$

$$\Rightarrow Y_3[3Y_2 - 3Y_1 - Y_3] > 0$$

Hence the solution. This example helps us to understand that under certain conditions, one can find estimators better than conventional estimators. ■

Problem 2.4 A simple random sample of size $n = n_1 + n_2$ with mean $\overline{y}$ is drawn from a finite population, and a simple random subsample of size n_1 is drawn from it with mean $\overline{y}_1$. Show that

(a) $V[\overline{y}_1 - \overline{y}_2] = S_y^2\left[\frac{1}{n_1} + \frac{1}{n_2}\right]$ where $\overline{y}_2$ is the mean of the remaining n_2 units in the sample,

(b) $V[\overline{y}_1 - \overline{y}] = S_y^2\left[\frac{1}{n_1} - \frac{1}{n}\right]$

(c) $\text{cov}(\overline{y}, \overline{y}_1 - \overline{y}) = 0$

Solution Since $\overline{y}_1$ is based on a subsample,

$$V(\overline{y}_1) = E_1V_2(\overline{y}_1) + V_1E_2(\overline{y}_1) \tag{2.17}$$

where E_1 is the unconditional expectation and E_2 the conditional expectation with respect to the subsample. Similarly V_1 is the unconditional variance and V_2 is the conditional variance with respect to the subsample.

It may be noted that $E_2[\overline{y}_1] = \overline{y}$ and $V_2[\overline{y}_1] = \frac{n - n_1}{nn_1} s_y^2$ (refer Remark 2.1).

Therefore $V_1E_2[\overline{y}_1] = \frac{N-n}{Nn} S_y^2$ and $E_1V_2[\overline{y}_1] = \frac{n-n_1}{nn_1} S_y^2$.

Substituting these expressions in (2.17) and doing the necessary simplification we get

$$V[\overline{y}_1] = \left(\frac{1}{n} - \frac{1}{N}\right) S_y^2 \tag{2.18}$$

Further

$$\begin{aligned}
\text{cov}(\overline{y}, \overline{y}_1) &= E[\overline{y}\,\overline{y}_1] - E[\overline{y}]E[\overline{y}_1] \\
&= E_1E_2[\overline{y}\,\overline{y}_1] - \overline{Y}E_1E_2[\overline{y}_1] \\
&= E_1[\overline{y}\,\overline{y}_1] - \overline{Y}\overline{Y} \\
&= V[\overline{y}] \\
&= \left[\frac{N-n}{Nn}\right] S_y^2
\end{aligned} \tag{2.19}$$

We know that

$$\begin{aligned}
\text{cov}(\overline{y}, \overline{y}_1 - \overline{y}) &= \text{cov}(\overline{y}, \overline{y}_1) - \text{cov}(\overline{y}, \overline{y}) \\
&= V[\overline{y}] - V[\overline{y}] \quad \text{(using 2.19)} \\
&= 0
\end{aligned}$$

This proves (c). ■

Note that

$$\begin{aligned}
V[\overline{y}_1 - \overline{y}] &= V[\overline{y}_1] + V[\overline{y}] - 2\,\text{cov}(\overline{y}, \overline{y}_1) \\
&= V[\overline{y}_1] + V[\overline{y}] - 2V[\overline{y}] \quad \text{(using 2.19)} \\
&= V[\overline{y}_1] - V[\overline{y}] \\
&= \left[\frac{N-n_1}{Nn_1}\right] S_y^2 - \left[\frac{N-n}{Nn}\right] S_y^2 = \left[\frac{1}{n_1} - \frac{1}{n}\right] S_y^2
\end{aligned}$$

This proves (b). ■

We know that $\bar{y} = \left[\dfrac{n_1\bar{y}_1 + n_2\bar{y}_2}{n}\right] \Rightarrow \bar{y}_2 = \dfrac{n\bar{y} - n_1\bar{y}_1}{n_2}$

Therefore $V[\bar{y}_1 - \bar{y}_2] = V\left[\bar{y}_1 - \dfrac{n}{n_2}\bar{y}_2 + \dfrac{n_1}{n_2}\bar{y}_2\right]$

$$= \frac{1}{n_2^2}V\left[n_2\bar{y}_1 - n\bar{y} + n_1\bar{y}_1\right] = \frac{1}{n_2^2}V[n(\bar{y}_1 - \bar{y})]$$

$$= \frac{n^2}{n_2^2}\left[\frac{1}{n_1} - \frac{1}{n}\right]S_y^2 \qquad \text{(using (b))}$$

$$= \frac{n^2}{n_2^2}\left[\frac{n - n_1}{nn_1}\right]S_y^2$$

$$= \frac{n}{n_2}\frac{1}{n_1}S_y^2 \qquad (\text{since } n_2 = n - n_1)$$

$$= \frac{n_1 + n_2}{n_1 n_2}S_y^2 = \left[\frac{1}{n_1} + \frac{1}{n_2}\right]S_y^2$$

This proves (a). ■

***Problem** 2.5* Suppose from a sample of n units selected with simple random sampling a subsample of n' units is selected with simple random sampling, duplicated and added to the original sample. Derive the expected value and the approximate sampling variance of $\bar{y}'$, the sample mean based on the $n+n'$ units. For what value of the fraction $\dfrac{n'}{n}$ does the efficiency of $\bar{y}'$ compared to that of $\bar{y}$ attains its minimum value?

Solution Denote by $\bar{y}^\circ$, the mean of the subsample. The sample mean based on $n+n'$ units can be written as

$$\bar{y}' = \frac{n\bar{y} + n'\bar{y}^\circ}{n+n'}$$

Since $\bar{y}'$ is based on the subsample,

$E[\bar{y}'] = E_1E_2[\bar{y}']$, where E_2 being expectation w.r.t. the subsample and E_1 the original sample.

Therefore $E[\bar{y}'] = E_1E_2\left[\dfrac{n\bar{y} + n'\bar{y}^\circ}{n+n'}\right] = \dfrac{nE_1E_2(\bar{y}) + n'E_1E_2(\bar{y}^\circ)}{n+n'}$

$$= \frac{n\bar{Y} + n'\bar{Y}}{n+n'} = \bar{Y}$$

Hence the combined mean is unbiased for the population mean.

$$V[\overline{y}'] = E_1V_2[\overline{y}'] + V_1E_2[\overline{y}']$$

$$= E_1V_2\left[\frac{n\overline{y}+n'\overline{y}^{\circ}}{n+n'}\right] + V_1E_2\left[\frac{n\overline{y}+n'\overline{y}^{\circ}}{n+n'}\right]$$

$$= \frac{1}{(n+n')^2}\{E_1V_2(n'\overline{y}^{\circ}) + V_1(n\overline{y}+n'\overline{y})\}$$

$$= \frac{1}{(n+n')^2}\left\{E_1\left[n'^2\frac{n-n'}{nn'}s_y^2\right] + (n+n')^2V_1(\overline{y})\right\}$$

$$= \left[\frac{n'}{n+n'}\right]^2\left[\frac{n-n'}{nn'}S_y^2\right] + \frac{N-n}{Nn}S_y^2$$

$$= \left[\frac{n'}{n+n'}\right]^2\left[\frac{n-n'}{nn'}S_y^2\right] + \frac{N-n}{Nn}S_y^2$$

$$= \left[\frac{n'}{n+n'}\right]^2\left[\frac{1}{n'}-\frac{1}{n}\right]S_y^2 + \frac{S_y^2}{n} \qquad \text{(approximately)}$$

$$= \left[\frac{\frac{n'}{n}}{1+\frac{n'}{n}}\right]^2\left[\frac{1}{n'}-\frac{1}{n}\right]S_y^2 + \left[\frac{1+\left(\frac{n'}{n}\right)}{1+\left(\frac{n'}{n}\right)}\right]^2\frac{S_y^2}{n}$$

$$= \left[3\left(\frac{n'}{n}\right)^2 + \frac{\frac{1}{n}}{\left[1+\frac{n'}{n}\right]^2}\right]S_y^2 = \frac{1+3\frac{n'}{n}}{\left[1+\left(\frac{n'}{n}\right)\right]^2}\frac{S_y^2}{n} \tag{2.20}$$

By Remark 2.1, $V(\overline{y}) = \left[\frac{N-n}{Nn}\right]S_y^2 = \frac{S_y^2}{n}$ (2.21)

Therefore by (2.20) and (2.21), the efficiency of $\overline{y}'$ as compared to $\overline{y}$ is

$$E = \frac{1+3\frac{n'}{n}}{\left[1+\frac{n'}{n}\right]^2} = \frac{1+3\theta}{(1+\theta)^2} \text{ where } \theta = \frac{n'}{n}.$$

Using calculus methods, it can be seen that E attains maximum at $\theta = \frac{2}{3}$.

Therefore the value of $\frac{n'}{n}$ for which the efficiency attains maximum is $\frac{2}{3}$. ■

Problem 2.6 Let y_i be the ith sample observation $(i = 1, 2, ..., N)$ in simple random sampling. Find the variance of y_i and the covariance of y_i and y_j $(i \neq j)$. Using these results derive the variance of the sample mean.

Solution

Claim : In simple random sampling, the probability of drawing the unit with label $r (r = 1, 2, ..., N)$ in the ith draw is same as the probability of drawing the unit with label r in the first draw.

Proof of the Claim

The probability of drawing the unit with label r in the first draw is $\frac{1}{N}$.

The probability of drawing the unit with label r in the ith draw is

$$\left[1-\frac{1}{N}\right]\left[1-\frac{1}{N-1}\right]\left[1-\frac{1}{N-2}\right]...\left[1-\frac{1}{N-i+2}\right]\left[1-\frac{1}{N-i+1}\right]$$

which on simplification reduces to $\frac{1}{N}$. Hence the claim.

Proceeding in the same way it can be seen that the probability of selecting the units with labels r and s in the ith and jth draws is same as the probability of drawing them in the first and second draws.

Therefore, we infer that y_i can take any one of the N values $Y_1, Y_2, ..., Y_N$ with equal probabilities $\frac{1}{N}$ and the product $y_i y_j$ can take the values $Y_1Y_2, Y_1Y_3, ..., Y_{N-1}Y_N$ with probabilities $\frac{1}{N(N-1)}$.

Hence we have

(i) $$E[y_i] = \frac{1}{N}\sum_{i=1}^{N} Y_i \tag{2.22}$$

(ii) $$E[y_i^2] = \frac{1}{N}\sum_{i=1}^{N} Y_i^2 \tag{2.23}$$

(iii) $$E[y_i y_j] = \frac{1}{N(N-1)}\sum_{i=1}^{N}\sum_{\substack{j=1 \\ i\neq j}}^{N} Y_i Y_j \tag{2.24}$$

From (2.22), we have $E[y_i] = \overline{Y}$.

Therefore $V[y_i] = \frac{1}{N}\sum_{j=1}^{N} Y_j^2 - \overline{Y}^2$ (refer (2.23))

$$= \frac{1}{N}\sum_{j=1}^{N}[Y_j - \overline{Y}]^2 = \frac{N-1}{N}S_y^2 \tag{2.25}$$

Using (2.24) and (2.25) we get

$$\text{cov}(y_i, y_j) = \frac{1}{N(N-1)} \sum_{i=1}^{N} \sum_{\substack{j=1 \\ i \neq j}}^{N} Y_i Y_j - \overline{Y}^2$$

$$= \frac{1}{N(N-1)} \left[\left(\sum_{k=1}^{N} Y_k \right)^2 - \sum_{k=1}^{N} Y_k^2 \right] - \overline{Y}^2$$

$$= \frac{1}{N(N-1)} \left[N^2 \overline{Y}^2 - \sum_{k=1}^{N} Y_k^2 \right] - \overline{Y}^2$$

$$= \frac{1}{(N-1)} \overline{Y}^2 - \frac{1}{N(N-1)} \sum_{k=1}^{N} Y_k^2$$

$$= -\frac{1}{N(N-1)} \left[\sum_{k=1}^{N} Y_k^2 - N\overline{Y}^2 \right] = -\frac{S_y^2}{N} \tag{2.26}$$

We know that

$$V[\overline{y}] = V\left[\frac{1}{n} \sum_{i=1}^{n} y_i \right] = \frac{1}{n^2} \left[\sum_{i=1}^{n} V(y_i) + 2 \sum_{\substack{i=1 \\ i<j}}^{n} \sum_{j=1}^{n} \text{cov}(y_i, y_j) \right]$$

$$= \frac{1}{n^2} \left[\frac{n(N-1)}{N} S_y^2 + 2 \frac{n(n-1)}{2} \left(\frac{-S_y^2}{N} \right) \right] \quad \text{(using (2.25) and (2.26))}$$

$$= \left[\frac{N-n}{Nn} \right] S_y^2$$

Hence the solution. ■

Problem 2.7 If the value of the population coefficient of variation $C = \frac{S_y}{\overline{Y}}$ is known at the estimation stage, is it possible to improve upon the estimator $\overline{y}$, the usual sample mean based on a sample of n units selected using simple random sampling? If so, give the improved estimator and obtain its efficiency, by comparing its mean square error with $V(\overline{y})$.

Solution Consider the estimator

$$\overline{y}_\lambda = \lambda \overline{y}$$

where λ is a constant.

The mean square error of the estimator $\overline{y}_\lambda$ is

$$MSE(\overline{y}_\lambda) = E[\lambda \overline{y} - \overline{Y}]^2$$

$$= E[\lambda(\overline{y} - \overline{Y}) + (\lambda - 1)\overline{Y}]^2$$

$$= \lambda^2 E(\bar{y}-\bar{Y})^2 + (\lambda-1)^2 \bar{Y}^2 + 2\lambda(\lambda-1)\bar{Y}E(\bar{y}-\bar{Y})$$

$$= \lambda^2 V(\bar{y}) + (\lambda-1)^2 \bar{Y}^2$$

$$= \frac{N-n}{Nn} S_y^2 + (\lambda-1)^2 \bar{Y}^2 \tag{2.27}$$

Using differential calculus methods, it can be seen that the above mean square error is minimum when

$$\lambda = \left[1 + \frac{N-n}{Nn} C^2\right]^{-1} \tag{2.28}$$

Therefore, the population mean can be estimated more precisely by using the estimator

$$\bar{y}_\lambda^\circ = \left[1 + \frac{N-n}{Nn} C^2\right]^{-1} \bar{y}$$

whenever the value of C is known.

Substituting (2.28) in (2.27) we get the minimum mean square error

$$M^* = \left[\frac{N-n}{Nn} S_y^2\right]\left[1 + \frac{N-n}{Nn} C^2\right]^{-1}$$

Therefore the relative efficiency of the improved estimator $\bar{y}_\lambda^\circ$ when compared to $\bar{y}$ is

$$\left[1 + \frac{N-n}{Nn} C^2\right]^{-1}$$

It may be noted that the above expression will always assume a value less than one. ■

Remark 2.2 We have pointed out, a simple random sample of size n is obtained by drawing n random numbers one by one without replacing and considering the units with the corresponding labels. If the random numbers are drawn with replacement and the units corresponding to the drawn numbers is treated as sample, we obtain what is known as a "Simple Random Sampling With Replacement " sample (*SRSWR*).

Problem 2.8 Show that in simple random sampling with replacement

(a) the sample mean $\bar{y}$ is unbiased for the population mean

(b) $V(\bar{y}) = \left[\frac{N-1}{Nn}\right] S_y^2$

Solution If $y_i, i = 1, 2, ..., N$ is the value of the unit drawn in the ith draw then y_i can take any one of the N values Y_i with probabilities $\frac{1}{N}$.

Therefore $E(y_i) = \sum_{j=1}^{N} Y_j \frac{1}{N} = \bar{Y}$ (2.29)

In the same way, we get

$$E(y_i^2) = \sum_{j=1}^{N} Y_j^2 \frac{1}{N} = \frac{1}{N}\sum_{j=1}^{N} Y_j^2$$

Hence $V(y_i) = \frac{1}{N}\sum_{j=1}^{N} Y_j^2 - \overline{Y}^2 = \frac{N-1}{N} S_y^2$ (2.30)

Since draws are independent $\operatorname{cov}(y_i, y_j) = 0$, we get

$$E(\overline{y}) = E\left[\frac{1}{n}\sum_{i=1}^{n} y_i\right] = \frac{1}{n} n\overline{Y} \text{ (using (2.29))}$$

$$= \overline{Y}$$

and

$$V(\overline{y}) = V\left[\frac{1}{n}\sum_{i=1}^{n} y_i\right] = \left[\frac{1}{n^2}\sum_{i=1}^{n} V(y_i)\right] = \frac{1}{n^2}(n)\frac{N-1}{N} S_y^2 = \frac{N-1}{Nn} S_y^2$$

Hence the solution. ■

Problem 2.9 A simple random sample of size 3 is drawn from a population of size N with replacement. As an estimator of $\overline{Y}$ we take $\overline{y}'$, the unweighted mean over the different units in the sample. Show that the average variance of $\overline{y}'$ is

$$\frac{(2N-1)(N-1)S_y^2}{6N^2}$$

Solution The sample drawn will contain 1,2 or 3 different units. Let P_1, P_2 and P_3 be the probabilities of the sample containing 1,2 and 3 different units respectively.

$$P_1 = \sum_{r=1}^{N} P \text{ (selecting } r\text{th unit in all the three draws)}$$

$$= N \frac{1}{N}\frac{1}{N}\frac{1}{N} = \frac{1}{N^2}$$

$P_2 = \sum_{r=1}^{N} P$ (selecting rth unit in draw 1 and a unit different from rth unit in the second and third draws) + $\sum_{r=1}^{N} P$ (selecting the rth unit in draw 2 and a unit different from rth unit in the first and third draws) + $\sum_{r=1}^{N} P$ (selecting the rth unit in draw 3 and a unit different from rth unit in the first and second draws).

$$= N\frac{1}{N}\frac{N-1}{N}\frac{N-1}{N} + N\frac{1}{N}\frac{N-1}{N}\frac{N-1}{N} + N\frac{1}{N}\frac{N-1}{N}\frac{N-1}{N}$$

$$= \frac{3(N-1)}{N^2}$$

$P_3 = \sum_{r=1}^{N} P$ (selecting rth unit in draw 1, a unit different from rth unit in the second draw and selecting in the third draw a unit different from units drawn in the first two draws)

$$= \frac{N(N-1)(N-2)}{N^3}$$

$$= \frac{(N-1)(N-2)}{N^2}$$

We know that the variance of the sample mean based on n distinct units is $\frac{N-n}{Nn} S_y^2$.

Therefore the average variance of $\bar{y}'$ is

$$\left(\frac{N-1}{N} S_y^2 \right)\frac{1}{N^2} + \left(\frac{N-2}{2N} S_y^2 \right)\frac{3(N-1)}{N^2} + \left(\frac{N-3}{3N} S_y^2 \right)\frac{(N-1)(N-2)}{N^3}.$$

which on simplification reduces to $\frac{(2N-1)(N-1)}{6N^2} S_y^2$.

Hence the solution. ■

Exercises

2.1 Derive $V(s_y^2)$ and $\mathrm{cov}(\bar{x}, s_x^2)$ in simple random without replacement under usual notations.

2.2 Let v denote the number of distinct units in a simple random sample drawn with replacement. Show that the sample mean based on the v distinct units is also unbiased for the population mean and derive its variance.

2.3 Suggest an unbiased estimator for the population proportion under simple random sampling without replacement and derive its variance and also obtain an estimator for the variance.

2.4 Suppose in a population of N units, NP units are known to have value zero. Obtain the relative efficiency of selecting n units from N units with simple random sampling with replacement as compared to selection of n units from the $N - NP$ non-zero units with simple random sampling with replacement in estimating the population mean.

2.5 A sample of size n is drawn from a population having N units by simple random sampling. A subsample of n_l units is drawn from the n units by simple random sampling. Let $\bar{y}_1$ denote the mean based on n_l units and $\bar{y}_2$ the mean based on $n - n_1$ units. Show that $w\bar{y}_1 + (1-w)\bar{y}_2$ is unbiased

for the population mean and derive its variance. Also derive the optimum value of w for which the variance attains minimum and the resulting estimator.

Chapter 3

Systematic Sampling Schemes

3.1 Introduction

In this chapter, a collection of sampling schemes called systematic sampling schemes which have several practical advantages are considered. In these schemes, instead of selecting n units at random, the sample units are decided by a single number chosen at random.

Consider a finite population of size N, the units of which are identified by the labels $1, 2, ..., N$ and ordered in ascending order according to their labels. Unless otherwise mentioned, it is assumed that the population size N is expressible as product of the sample size n and some positive integer k, which is known as the reciprocal of the sampling fraction or sampling interval.

In the following section we shall describe the most popular linear systematic sampling scheme abbreviated as LSS.

3.2 Linear Systematic Sampling

A Linear Systematic Sample (LSS) of size n is drawn by using the following procedure :

Draw at random a number less than or equal to k, say r. Starting from the rth unit in the population, every kth unit is selected till a sample of size n is obtained.

For example, when N=24, n=6 and k=4, the four possible linear systematic samples are :

Sample Number	Random Start	Sampled units
1	1	1, 5, 9, 13
2	2	2, 6, 10, 14
3	3	3, 7, 11, 15
4	4	4, 8, 12, 16

The linear systematic sampling scheme described above can be regarded as dividing the population of N units into k mutually exclusive and exhaustive groups $\{S_1, S_2, ..., S_k\}$ of n units each and choosing one of them at random where the units in the rth group are given by

$$S_r = \{r, r+k, ..., r+(n-1)k\}, r = 1, 2, ..., k$$

The following theorem gives an unbiased estimator for the population total and its variance under LSS.

Theorem 3.1 An unbiased estimator for the population total Y under LSS corresponding to the random start r is given by $\hat{Y}_{LSS} = \frac{N}{n}\sum_{j=1}^{n} Y_{r+(j-1)k}$ and its variance is $V(\hat{Y}_{LSS}) = \frac{1}{k}\sum_{r=1}^{k}[\hat{Y}_r - Y]^2$ where $\hat{Y}_r$ is the value of $\hat{Y}_{LSS}$ corresponding to the random start r.

Proof Note that the estimator $\hat{Y}_{LSS}$ can take any one of the k values $\hat{Y}_r, r = 1, 2, ..., k$ with equal probabilities $\frac{1}{k}$.

Therefore

$$E[\hat{Y}_{LSS}] = \sum_{r=1}^{k} \hat{Y}_r\left(\frac{1}{k}\right) = \frac{1}{k}\sum_{r=1}^{k}\frac{N}{n}\sum_{j=1}^{n} Y_{r+(j-1)k}$$

$$= \frac{N}{nk}\sum_{r=1}^{k}\sum_{j=1}^{n} Y_{r+(j-1)k} = \sum_{i=1}^{N} Y_i \tag{3.1}$$

Hence $\hat{Y}_{LSS}$ is unbiased for the population total Y.

Since the estimator $\hat{Y}_{LSS}$ can take any one of the k values $\hat{Y}_r, r = 1, 2, ..., k$ with equal probabilities $\frac{1}{k}$ and it is unbiased for Y,

$$V(\hat{Y}_{LSS}) = E[\hat{Y}_{LSS} - Y]^2$$

$$= \sum_{r=1}^{k}[\hat{Y}_r - Y]^2\left(\frac{1}{k}\right)$$

$$= \frac{1}{k}\sum_{r=1}^{k}[\hat{Y}_r - Y]^2 \tag{3.2}$$

Hence the proof. ■

Apart from operational convenience, the linear systematic sampling has an added advantage over simple random sampling namely, the simple expansion estimator defined in the above theorem is more precise than the corresponding estimator in simple random sampling for populations exhibiting linear trend. That is, if the values $Y_1, Y_2, ..., Y_N$ of the units with labels $1, 2, ..., N$ are modeled by $Y_i = \alpha + \beta i, i = 1, 2, ..., N$ then systematic sampling is more efficient than simple random sampling when the simple expansion estimator is used for estimating the population total. This is proved in the following theorem. Before the theorem is stated, we shall give a frequently used identity meant for populations possessing linear trend.

Identity For populations modeled by $Y_i = \alpha + \beta i, i = 1, 2, ..., N$

$$\hat{Y}_r - Y = N\beta\left[r - \frac{(k+1)}{2}\right] \tag{3.3}$$

where $\hat{Y}_r$ is as defined in Theorem 3.1.

Proof : Note that when $Y_i = \alpha + \beta i, i = 1, 2, ..., N$, we have

$$\hat{Y}_r = \frac{N}{n}\sum_{j=1}^{n} Y_{r+(j-1)k} = \frac{N}{n}\sum_{j=1}^{n}\{\alpha + \beta[r + (j-1)k]\}$$

$$= \frac{N}{n}\left[n\alpha + \beta nr + \beta[k\frac{n(n-1)}{2}]\right]$$

$$= N\left[\alpha + \beta r + \beta[k\frac{(n-1)}{2}]\right] \tag{3.4}$$

and $$Y = \sum_{i=1}^{N} Y_i$$

$$= \sum_{i=1}^{N}[\alpha + \beta i] = N\alpha + \beta\left[\frac{N(N+1)}{2}\right] \tag{3.5}$$

Using (3.3) and (3.4) we get

$$\hat{Y}_r - Y = N\left[\alpha + \beta r + \beta\left[\frac{k(n-1)}{2}\right] - \alpha - \beta\left[\frac{nk+1}{2}\right]\right]$$

$$= N\beta\left[r + \frac{nk - k - nk - 1}{2}\right]$$

$$= N\beta\left[r - \frac{(k+1)}{2}\right]$$

Hence the identity given in (3.3) holds good for all r, $r = 1, 2, ..., k$. ■

Theorem 3.2 For populations possessing linear trend, $V(\hat{Y}_{LSS}) < V(\hat{Y}_{srs})$ where $\hat{Y}_{LSS}$ and $\hat{Y}_{srs}$ are the conventional expansion estimators under linear systematic sampling and simple random sampling, respectively.

Proof We know that under simple random sampling

$$V(\hat{Y}_{srs}) = \frac{N^2(N-n)}{Nn}\frac{1}{N-1}\sum_{i=1}^{N}[Y_i - \bar{Y}]^2 \tag{3.6}$$

Note that if $Y_i = \alpha + \beta i, i = 1, 2, ..., N$, then $\bar{Y} = \alpha + \beta\frac{(N+1)}{2}$ (3.7)

Therefore, for $i = 1, 2, ..., N$

$$[Y_i - \overline{Y}]^2 = \left[\alpha + \beta i - \alpha - \beta\left(\frac{N+1}{2}\right)\right]^2$$

$$= \beta^2\left[i - \left(\frac{N+1}{2}\right)\right]^2$$

$$= \beta^2\left[i^2 + \frac{(N+1)^2}{4} - i(N+1)\right]$$

Hence $\sum_{i=1}^{N}[Y_i - \overline{Y}]^2 = \beta^2\left[\frac{N(N+1)(2N+1)}{6} + \frac{N(N+1)^2}{4} - \frac{N(N+1)^2}{2}\right]$

$$= \beta^2\left[\frac{N(N+1)(N-1)}{12}\right]$$

Substituting this in (3.6) we get

$$V(\hat{Y}_{srs}) = \frac{N^2(N-n)}{Nn}\frac{1}{N-1}\beta^2\left[\frac{N(N+1)(N-1)}{12}\right]$$

$$= \frac{N^2\beta^2(k-1)(nk+1)}{12} \quad \text{(using } N=nk\text{)} \tag{3.8}$$

On using identity given in (3.3), we get

$$\sum_{r=1}^{k}[\hat{Y}_r - Y]^2 = \sum_{r=1}^{k} N^2\beta^2\left[r - \frac{(k+1)}{2}\right]^2$$

$$= N^2\beta^2\sum_{r=1}^{k}\left[r^2 + \frac{(k+1)^2}{4} - i(k+1)\right]$$

$$= N^2\beta^2\left[\frac{k(k+1)(2k+1)}{6} + \frac{k(k+1)^2}{4} - \frac{k(k+1)^2}{2}\right]$$

$$= \frac{N^2\beta^2 k(k^2-1)}{12}$$

Therefore $V(\hat{Y}_{LSS}) = \dfrac{N^2\beta^2(k^2-1)}{12}$ (3.9)

Thus using (3.8) and (3.9) we get

$$V(\hat{Y}_{srs}) - V(\hat{Y}_{LSS}) = \frac{N^2\beta^2(k-1)(nk+1-k-1)}{12}$$

$$= \frac{N^2\beta^2 k(n-1)(k-1)}{12}$$

Since the right hand side of the above expression is positive for all values of n greater than one, the result follows. ■

Yates Corrected Estimator

In Theorem 3.2, it has been proved that linear systematic sampling is more precise than simple random sampling in the presence of linear trend. Yates (1948) suggested an estimator that coincides with the population mean under linear systematic sampling for populations possessing linear trend. The details are furnished below:

When the rth group S_r is drawn as sample, the first and last units in the sample are corrected by the weights λ_1 and λ_2 respectively (that is, instead of using Y_r and $Y_{r+(n-1)k}$ in the estimator, the corrected values namely $\lambda_1 Y_r$ and $\lambda_2 Y_{r+(n-1)k}$ will be used) and the sample mean is taken as an estimator for the population mean, where the weights λ_1 and λ_2 are selected so that the corrected mean coincides with the population mean in the presence of linear trend. That is, the corrected mean

$$\overline{y}_c = \frac{1}{n}\left[\lambda_1 Y_r + \sum_{j=2}^{n-1} Y_{r+(j-1)k} + \lambda_2 Y_{r+(n-1)k}\right]$$

is equated to the population mean $\overline{Y}$ after substituting, $Y_i = \alpha + \beta i, i = 1, 2, \ldots, N$ to get

$$\frac{1}{n}\left[\begin{array}{l}\lambda_1(\alpha+\beta r)+\sum_{j=2}^{n-1}[\alpha+\beta[r+(j-1)k]]+\\ \lambda_2[\alpha+\beta[r+(n-1)k]]\end{array}\right] = \alpha + \frac{\beta(N+1)}{2} \tag{3.10}$$

Comparing the coefficients of α in (3.10) we get

$$\frac{1}{n}[\lambda_1 + \lambda_2 + n - 2] = 1$$

Therefore $\lambda_1 + \lambda_2 = 2$ (3.11)

Again comparing the coefficient of β in (3.10) we get

$$\frac{1}{n}[\lambda_1 r + \lambda_2[r+(n-1)k] + (n-2)r + \frac{(n-1)(n-2)}{2}k\ ] = \frac{N+1}{2}$$

$$2[\lambda_1 r + \lambda_2[r+(n-1)k] + (n-2)r + \frac{(n-1)(n-2)}{2}k\ ] = n(N+1)$$

$$[2\lambda_1 r + 2(2-\lambda_1)(n-1)k + 2(n-2)r + (n-1)(n-2)k\ ] = n(N+1)$$

(using (3.11))

Solving for λ_1 we get $\lambda_1 = 1 + \dfrac{n(2r-k-1)}{2(n-1)k}$ (3.12)

Using (3.12) in (3.13), we find that

$$\lambda_2 = 1 - \frac{n(2r-k-1)}{2(n-1)k} \tag{3.13}$$

When the above obtained values of λ_1 and λ_2 are used in the Yates corrected estimator, we get

$$\bar{y}_c = \frac{1}{N}\left[Y_r + \frac{(2r-k-1)}{2(n-1)k}[Y_r - Y_{r+(n-1)k}] \right] \tag{3.14}$$

Therefore the estimator

$$\hat{Y}_c = \left[Y_r + \frac{(2r-k-1)}{2(n-1)k}[Y_r - Y_{r+(n-1)k}] \right]$$

estimates the population total without any error. Since the estimator coincides with the parameter value, it has mean square error zero.

3.3 Schemes for Populations with Linear Trend

In the previous section, we have seen a method in which the corrected expansion estimator coincides with the population total in the presence of linear trend. However, instead of correcting the estimator, many have suggested alternative sampling schemes which are best suitable for populations with linear trend. Three such schemes are presented in this section.

(i) Centered Systematic Sampling (Madow, 1953)

As in the case of linear systematic sampling, in centered systematic sampling also the population units are divided into k groups $S_1, S_2, ..., S_k$ of n units each, where $S_r = \{r, r+k, ..., r+(n-1)k\}, r = 1, 2, ..., k$.

If the sampling interval k is odd then the middlemost group namely $S_{(k+1)/2}$ is selected as sample with probability one. On the other hand, one of the middlemost groups, namely $S_{k/2}$ or $S_{(k+2)/2}$ will be randomly selected as sample.

To estimate the population total, one can use the expansion estimator as in the case of linear systematic sampling. If $\hat{Y}_{CSS}$ is the estimator of the population total under centered systematic sampling, then

(i) when k is odd, $\hat{Y}_{CSS} = \hat{Y}_{(k+1)/2}$ with probability one

(ii) when k is even, $\hat{Y}_{CSS} = \begin{cases} \hat{Y}_{k/2} & \text{with probability } 1/2 \\ \hat{Y}_{(k+2)/2} & \text{with probability } 1/2 \end{cases}$

It may be noted that in both the cases $\hat{Y}_{CSS}$ is not unbiased for Y. However, for populations with linear trend, it has same desirable properties as shown in the following theorem.

Theorem 3.3 For populations satisfying $Y_i = \alpha + \beta i, i = 1, 2, ..., N$,

(i) when k is odd, $\hat{Y}_{CSS} = Y$ and $MSE(\hat{Y}_{CSS}) = 0$ and

(ii) when k is even, $E(\hat{Y}_{CSS}) = Y$ and $MSE(\hat{Y}_{CSS}) = \frac{\beta^2}{4}$

Proof For populations with linear trend, we have seen in (3.3)

$$\hat{Y}_r - Y = N\beta\left[r - \frac{(k+1)}{2}\right], r = 1, 2, \ldots, k$$

Therefore (i) $\hat{Y}_{(k+1)/2} - Y = N\beta\left[\frac{(k+1)}{2} - \frac{(k+1)}{2}\right] = 0$ (3.15)

(ii) $\hat{Y}_{k/2} - Y = N\beta\left[\frac{k}{2} - \frac{(k+1)}{2}\right] = -\frac{N\beta}{2}$ (3.16)

and (iii) $\hat{Y}_{(k+2)/2} - Y = N\beta\left[\frac{k+2}{2} - \frac{(k+1)}{2}\right] = \frac{N\beta}{2}$ (3.17)

Hence when k is odd $MSE(\hat{Y}_{CSS}) = [\ \hat{Y}_{(k+1)/2} - Y\ ]^2 = 0$ (By (3.15))

and when k is even $MSE(\hat{Y}_{CSS}) = \frac{1}{2}[\ \hat{Y}_{k/2} - Y\ ]^2 + \frac{1}{2}[\ \hat{Y}_{(k+2)/2} - Y\ ]^2$

$$= \frac{1}{2}\left[\frac{N^2\beta^2}{4} + \frac{N^2\beta^2}{4}\right]$$

$$= \frac{N^2\beta^2}{4}$$

Thus we have proved the theorem. ■

The centered systematic sampling described above is devoid of randomisation. Hence the results based on a centered systematic sample are likely to be unreliable particularly when the assumption regarding the presence of linear trend is violated. Hence it is desirable to develop a sampling method free from such limitation. In the following part of this section, one such scheme developed by Sethi(1962) is presented.

(ii) Balanced Systematic Sampling (Sethi, 1962)

Under Balanced Systematic Sampling (BSS), the population units are divided into $\frac{n}{2}$ groups (assuming the sample size n is even) of $2k$ units each and a pair of units equidistant from the end points are selected from each group. This is achieved by using the following procedure:

A random number r is selected from 1 to k and units with labels r and $2k - r + 1$ will be selected from the first group and thereafter from the remaining $\frac{n}{2} - 1$ groups, the corresponding pairs of elements will be selected. For example, when N=24 and n=6, the population units are divided into $\frac{6}{2} = 3$ groups of (2)(4)=8 units each as follows:

1	2	3	4	5	6	7	8
9	10	11	12	13	14	15	16
17	18	19	20	21	22	23	24

The four possible balanced systematic samples are listed below:

$$s_1 = \{1, 9, 17, 8, 6, 24\},\ s_2 = \{2, 10, 18, 7, 15, 23\},$$
$$s_3 = \{3, 11, 19, 6, 14, 22\},\ s_4 = \{5, 13, 21, 4, 12, 20\}$$

Thus the balanced systematic sample of size n corresponding to the random start r is given by the units with labels

$$\{r + 2jk, 2(j+1)k - r + 1\},\ j = 0, 1, 2, \ldots, \frac{n}{2} - 1$$

When the sample size n is odd, the balanced systematic sample of size n corresponding to the random start r is given by the units with labels

$$\{r + 2jk, 2(j+1)k - r + 1\} \cup \{r + (n-1)k\},\ j = 0, 1, 2, \ldots, \frac{n-3}{2}$$

Theorem 3.4 Under balanced systematic sampling, the conventional expansion estimator is unbiased for the population total.

Proof **Case 1 "*n* even"**

The expansion estimator $\hat{Y}_{BL}$ corresponding to the random start r can take any one of the k values

$$\hat{Y}_{BL}^{(r)} = \frac{N}{n} \sum_{j=0}^{(n-2)/2} [Y_{r+2jk} + Y_{2(j+1)k-r+1}],\ r = 1, 2, \ldots, k$$

with equal probabilities $\frac{1}{k}$.

Therefore $E(\hat{Y}_{BL}) = \frac{1}{k} \sum_{r=1}^{k} \hat{Y}_{BL}^{(r)}$

$$= \frac{1}{k} \sum_{r=1}^{k} \frac{N}{n} \sum_{j=0}^{(n-2)/2} \{Y_{r+2jk} + Y_{2(j+1)k-r+1}\}$$

$$= \frac{N}{nk} \sum_{r=1}^{k} \sum_{j=0}^{(n-2)/2} \{Y_{r+2jk} + Y_{2(j+1)k-r+1}\}$$

$$= \sum_{i=1}^{N} Y_i \quad \left(\text{since } \bigcup_{r=1}^{k} s_r = S \text{ and } s_r \cap s_t = \phi \text{ for } r \neq t\right)$$

Hence $\hat{Y}_{BL}$ is unbiased for the population total Y.

Case 2 "*n* odd"

In this case, $\hat{Y}_{BL}$ can take any one of the k values

$$\hat{Y}_{BL}^{(r)} = \frac{N}{n} \left[\sum_{j=0}^{(n-3)/2} [Y_{r+2jk} + Y_{2(j+1)k-r+1}] + Y_{r+(n-1)k} \right],\ r = 1, 2, \ldots, k$$

with equal probabilities $\frac{1}{k}$.

Therefore $E(\hat{Y}_{BL}) = \frac{1}{k}\sum_{r=1}^{k}\hat{Y}_{BL}^{(r)}$

$$= \frac{1}{k}\sum_{r=1}^{k}\frac{N}{n}\left\{\sum_{j=0}^{(n-3)/2}[Y_{r+2jk} + Y_{2(j+1)k-r+1}] + Y_{r+(n-1)k}\right\}$$

$$= Y \quad (\text{since } \bigcup_{r=1}^{k} s_r = S \text{ and } s_r \cap s_t = \phi \text{ for } r \neq t)$$

Hence in this case also, $\hat{Y}_{BL}$ is unbiased for the population total. ■

Thus from the above theorem we infer that the conventional estimator is unbiased for the population total in balanced systematic sampling. It may also be noted that the variance of the estimator is

$$V(\hat{Y}_{BL}) = \frac{1}{k}\sum_{r=1}^{k}[\hat{Y}_{BL}^{(r)} - Y]^2$$

where $\hat{Y}_{BL}^{(r)}$ is as defined in the previous theorem.

Theorem 3.5 When $Y_i = \alpha + \beta i, i = 1,2,\ldots,N$,

$$V(\hat{Y}_{BL}) = \begin{cases} 0 & \text{when n is even} \\ \dfrac{\beta^2 k^2 (k^2-1)}{12} & \text{when n is odd} \end{cases}$$

Proof For $r = 1, 2, \ldots, k$, when n is even

$$\hat{Y}_{BL}^{(r)} = \frac{N}{n}\sum_{j=0}^{(n-2)/2}[Y_{r+2jk} + Y_{2(j+1)k-r+1}]$$

$$= \frac{N}{n}\sum_{j=0}^{(n-2)/2}\{\alpha + \beta(r+2jk) + \alpha + \beta[2(j+1)k - r + 1]\}$$

$$= \frac{N}{n}\sum_{j=0}^{(n-2)/2}\{2\alpha + \beta[r + 2jk + 2(j+1)k - r + 1]\}$$

$$= \frac{N}{n}\left\{\frac{n}{2}\{2\alpha + \beta(2k+1)\} + 4\beta k\frac{n(n-2)}{8}\right\}$$

$$= N\left[\alpha + \frac{\beta(nk+1)}{2}\right] = Y$$

Thus we have $\hat{Y}_{BL}^{(r)} = Y$ for all $r = 1, 2, \ldots, k$

Therefore $V(\hat{Y}_{BL}) = 0$.

For $r = 1, 2, \ldots, k$, when n is odd

$$\hat{Y}_{BL}^{(r)} = \frac{N}{n}\left\{ \sum_{j=0}^{(n-3)/2} [Y_{r+2jk} + Y_{2(j+1)k-r+1}] + Y_{r+(n-1)k} \right\}$$

$$= \frac{N}{n}\left\{ \sum_{j=0}^{(n-3)/2} [2\alpha + \beta(4jk + 2k + 1)] + \{\alpha + \beta[r + (n-1)k\} \right\}$$

$$= \frac{N}{n}\left[n\alpha + \beta\left\{ \frac{(n-1)(2k+1)}{2} + \frac{(n-1)(n-3)k}{2} + r + (n-1)k \right\}\right]$$

$$= \frac{N}{n}\left[n\alpha + \beta\left\{ \frac{(n-1)(nk+k+1)}{2} + r \right\}\right] \tag{3.18}$$

Further we know that for populations having linear trend

$$Y = N\alpha + \beta\frac{(nk)(nk+1)}{2} \tag{3.19}$$

Using (3.18) and (3.19) we get, for $r = 1, 2, ..., k$

$$\hat{Y}_{BL}^{(r)} - Y = \frac{\beta N}{n}\left[r - \frac{k+1}{2} \right]$$

Squaring both sides and summing with respect to k we get

$$\frac{1}{k}\sum_{r=1}^{k}\left[\hat{Y}_{BL}^{(r)} - Y\right]^2 = \frac{\beta^2 N^2}{n^2 k}\sum_{r=1}^{k}\left[r - \frac{k+1}{2}\right]^2$$

$$= \frac{\beta^2 N^2}{n^2 k}\left[\sum_{r=1}^{k} r^2 + \frac{k(k+1)^2}{4} - 2\frac{k+1}{2}\frac{k(k+1)}{2} \right]$$

$$= \frac{\beta^2 N^2}{n^2 k}\left[\frac{k(k+1)(2k+1)}{6} - \frac{k(k+1)^2}{4} \right]$$

$$= \frac{\beta^2 (k^2 - 1)k^2}{12}$$

Hence the theorem. ■

(iii) Modified Systematic Sampling (Singh, Jindal and Garg, 1965)

The modified systematic sampling is another scheme meant for populations exhibiting linear trend. A sample of size n is drawn by selecting pairs of units equidistant from both the ends of the population in a systematic manner. The details are furnished below.

As in the case of linear and balanced systematic sampling here also a random number r is selected from 1 to k. When the sample size n is even, the sample corresponding to the random start r ($r = 1, 2, ..., k$) is given by the set of units with labels

$$s_r = \{r + jk, N - r - jk + 1\},\ j = 0, 1, ..., \frac{n-2}{2}$$

When the sample size n is odd, the sample corresponding to the random start r ($r = 1, 2, ..., k$) is given by the set of units with labels

$$s_r = \{r + jk, N - r - jk + 1\} \bigcup_{r=1}^{k} \{r + \frac{(n-1)k}{2}, j = 0, 1, ..., \frac{n-3}{2}$$

For example, when N=16, n=4 and k=4 the four possible modified systematic samples are

$$s_1 = \{1, 5, 12, 16\}, s_2 = \{2, 6, 11, 15\}, s_3 = \{3, 7, 10, 14\}, s_4 = \{4, 8, 9, 13\}$$

It is interesting to note that the theorems which we proved in the previous section for balanced systematic sampling are true even under modified systematic sampling.

Theorem 3.6 Under Modified Systematic sampling, the conventional expansion estimator is unbiased for the population total.

Proof of this theorem is left as exercise.

Theorem 3.7 Under Modified Systematic sampling,

$$V(\hat{Y}_{MOD}) = \begin{cases} 0 \text{ if } n \text{ is even} \\ \dfrac{\beta^2 k^2 (k^2 - 1)}{12} \text{ if } n \text{ is odd} \end{cases}$$

when $Y_i = \alpha + \beta i, , i = 1, 2, ..., N$, where $\hat{Y}_{MOD}$ is the conventional estimator under Modified Systematic Sampling.

Proof of this theorem is also left as an exercise.

3.4 Autocorrelated Populations

Generally, it is reasonable to expect the units which are nearer to each other possess similar characteristics. This property can be represented using a statistical model assuming that the observations Y_i and Y_j are correlated, the correlation being a function of the distance between the labels of the units which decreases as the distance increases. Such populations are called autocorrelated populations and the graph of the correlation coefficient ρ_u, between observations separated by u units, as a function of u is called "correlogram".

Because of the finite nature of the population the correlogram will not be smooth and it will be difficult to study the relative performances of various sampling schemes for a single finite population. But it is easy on the average over a series of finite populations drawn from an infinite super population to which the model applies. A model which is suitable for populations possessing autocorrelatedness is described below.

Model The population values are assumed to be the realized values of N random variables satisfying the following conditions :

$$E_M[Y_i] = \mu \, , E_M[Y_i - \mu]^2 = \sigma^2$$

and $E_M[Y_i - \mu][Y_{i+u} - \mu] = \rho_u \sigma^2$ where $\rho_u \geq \rho_v$ whenever $u < v$.

The subscript M is used to denote the fact that the expectations are with respect to the superpopulation model .The following theorem gives the average variance of the expansion estimator under simple random sampling and linear systematic sampling.

Theorem 3.8 Under the super population model described above

$$E_M V(\hat{Y}_{srs}) = \frac{\sigma^2 (k-1) N^2}{nk} \left[1 - \frac{2}{N(N-1)} \sum_{u=1}^{N-1} (N-u) \rho_u \right]$$

$$E_M V(\hat{Y}_{LSS}) = \frac{\sigma^2 (k-1) N^2}{nk} \left\{ \left[1 - \frac{2}{nk(k-1)} \sum_{u=1}^{nk-1} (nk-u) \rho_u \right] + \left[\frac{2k}{n(k-1)} \sum_{u=1}^{n-1} (n-u) \rho_{ku} \right] \right\}$$

Proof We know that $V(\hat{Y}_{srs}) = \dfrac{N^2(N-n)}{Nn} \dfrac{1}{N-1} \sum_{i=1}^{N} [Y_i - \overline{Y}]^2$ (3.20)

Note that $\sum_{i=1}^{N} [Y_i - \mu]^2 = \sum_{i=1}^{N} [Y_i - \overline{Y} + \overline{Y} - \mu]^2$

$$= \sum_{i=1}^{N} [Y_i - \overline{Y}]^2 + N[\overline{Y} - \mu]^2$$

Therefore $\sum_{i=1}^{N} [Y_i - \overline{Y}]^2 = \sum_{i=1}^{N} [Y_i - \mu]^2 - N[\overline{Y} - \mu]^2$

$$= \sum_{i=1}^{N} [Y_i - \mu]^2 - N \left[\frac{1}{N} \sum_{i=1}^{N} [Y_i - \mu] \right]^2$$

$$= \sum_{i=1}^{N} [Y_i - \mu]^2 - \frac{1}{N} \left[\sum_{i=1}^{N} [Y_i - \mu]^2 + 2 \sum_{i<j}^{N} [Y_i - \mu][Y_j - \mu] \right]$$

$$= \frac{N-1}{N} \sum_{i=1}^{N} [Y_i - \mu]^2 - \frac{2}{N} \sum_{u=1}^{N-1} \sum_{i=1}^{N-u} [Y_i - \mu][Y_{i+u} - \mu]$$

Taking expectations on both the sides with respect to the model we obtain

$$E_M \left[\sum_{i=1}^{N} [Y_i - \overline{Y}]^2 \right] = (N-1) \sigma^2 \left[1 - \frac{2}{N(N-1)} \sum_{u=1}^{N-1} (N-u) \rho_u \right] \tag{3.21}$$

Using (3.20) and (3.21) we get

$$E_M V(\hat{Y}_{srs}) = \frac{N^2(N-n)\sigma^2(N-1)}{Nn(N-1)}\left[1 - \frac{2}{N(N-1)}\sum_{u=1}^{N-1}(N-u)\rho_u\right]$$

$$= \frac{\sigma^2(k-1)N^2}{nk}\left[1 - \frac{2}{N(N-1)}\sum_{u=1}^{N-1}(N-u)\rho_u\right]$$

Thus we have obtained the average variance of the conventional estimator under simple random sampling with respect to the autocorrelated model described earlier.

We know that $V(\hat{Y}_{LSS}) = \frac{1}{k}\sum_{r=1}^{k}[\hat{Y}_r - Y]^2$

$$= \frac{N^2}{k}\sum_{r=1}^{k}[\hat{\bar{Y}}_r - \bar{Y}]^2$$

where $\hat{\bar{Y}}_r = \frac{1}{n}\sum_{j=1}^{n} Y_{r+(j-1)k}$ and $\bar{Y} = \frac{1}{k}\sum_{r=1}^{k}\hat{\bar{Y}}_r$

Note that $\sum_{i=1}^{N}[Y_i - \bar{Y}]^2 = \sum_{r=1}^{k}\sum_{j=1}^{n}[Y_{r+(j-1)k} - \bar{Y}]^2$

$$= \sum_{r=1}^{k}\sum_{j=1}^{n}[Y_{r+(j-1)k} - \hat{\bar{Y}}_r + \hat{\bar{Y}}_r - \bar{Y}]^2$$

$$= \sum_{r=1}^{k}\sum_{j=1}^{n}[Y_{r+(j-1)k} - \hat{\bar{Y}}_r]^2 + n\sum_{r=1}^{k}[\hat{\bar{Y}}_r - \bar{Y}]^2$$

Therefore $$\sum_{r=1}^{k}[\hat{\bar{Y}}_r - \bar{Y}]^2 = \frac{1}{n}\sum_{i=1}^{N}[Y_i - \bar{Y}]^2 - \frac{1}{n}\sum_{r=1}^{k}\sum_{j=1}^{n}[Y_{r+(j-1)k} \quad \hat{\bar{v}}\;]^2 \tag{3.23}$$

On using (3.21) we get

$$E_M\left[\sum_{j=1}^{n}[Y_{r+(j-1)k} - \hat{\bar{Y}}_r]^2\right] = (n-1)\sigma^2\left[1 - \frac{2}{n(n-1)}\sum_{u=1}^{n-1}(n-u)\rho_{ku}\right] \tag{3.24}$$

Using (3.21) and (3.24) in (3.23) we obtain

$$E_M \sum_{r=1}^{k} [\hat{\bar{Y}}_r - \bar{Y}]^2 = \frac{1}{n}\left[(N-1)\sigma^2\right]\left\{1 - \frac{2}{N(N-1)}\sum_{u=1}^{N-1}(N-u)\rho_u\right\} -$$

$$\frac{(n-1)k\sigma^2}{n}\left\{1 - \frac{2}{n(n-1)}\sum_{u=1}^{n-1}(n-u)\rho_{ku}\right\}$$

$$= \frac{\sigma^2}{n}\left\{(k-1) - \frac{2}{nk}\sum_{u=1}^{N}(N-u)\rho_u + \frac{2k}{n}\sum_{u=1}^{n-1}(n-u)\rho_{ku}\right\}$$

$$= \frac{\sigma^2(k-1)}{n}\left\{1 - \frac{2}{nk(k-1)}\sum_{u=1}^{nk}(nk-u)\rho_u + \frac{2k}{n(k-1)}\sum_{u=1}^{n-1}(n-u)\rho_{ku}\right\}$$

Substituting this in (3.22) we get the average variance of the conventional estimator in linear systematic sampling. ■

A comparison between these two average variances is given in Chapter 5.

3.5 Estimation of Variance

In Linear systematic sampling the second order inclusion probabilities are not positive for all pairs of units in the population. This makes unbiased estimation of the variance of the estimator impossible. In the absence of a proper estimate for the variance, several ad hoc procedures are being followed to estimate the variance of the conventional estimator.

One of the methods is to treat the systematic sample as a simple random sample of size n units and estimate the variance by

$$\frac{N^2(N-n)}{Nn}\frac{1}{n-1}\sum_{i=1}^{n}[y_i - \bar{y}]^2$$

where y_i is the y-value of the ith unit in the sample and $\bar{y}$ is the sample mean. It may be noted that the above estimator is not unbiased for the variance of conventional estimator under linear systematic sampling.

The second approach is to treat systematic sampling as a process of grouping the N population units into $\frac{n}{2}$ groups of $2k$ units each and selecting two units from each group in a systematic manner. In this case the population total can be estimated by

$$N\frac{2}{n}\sum_{i=1}^{n/2}\frac{y_{2i}+y_{2i-1}}{2} \tag{3.22}$$

Assuming that the two units have been selected with simple random sampling without replacement from the $2k$ units in the ith group, an unbiased estimator of the variance of the ith term in the brackets on the right hand side of (3.22) will be given by

$$\frac{(k-1)}{k}\left\{\frac{y_{2i}-y_{2i-1}}{2}\right\}^2$$

Hence an unbiased estimator of the variance of the estimator given in (3.22) is

$$N^2\frac{N-n}{Nn^2}\sum_{i=1}^{n/2}\left\{\frac{y_{2i}+y_{2i-1}}{2}\right\}^2$$

An alternative variance estimator based on the same principles as those considered above, which takes into account successive differences of sample values is given by

$$N^2\frac{N-n}{Nn}\sum_{i=1}^{n-1}\frac{\{y_{i+1}-y_i\}^2}{2(n-1)}$$

Singh and Singh (1977) proposed a new type of systematic sampling which facilitates the estimation of variance under certain conditions. The scheme suggested by them is described below.

(i) Select a random number from 1 to N

(ii) Starting with r select u continuous units and the remaining $n-u=v$ units with interval d, where u(less than or equal to n) and d are predetermined.

They have proved that if $u+vd$ is less than or equal to N then the above sampling scheme will yield distinct units and the second order inclusion probabilities are positive if (a) d is less than or equal to u and (b) $u+vd$ is greater than or equal to $(N/2)+1$. When these two conditions are satisfied it is possible to estimate the variance of the conventional estimator. They have observed that in situations where usual systematic sampling performs better than simple random sampling the suggested procedure also leads to similar results and for some situations, it provides better results than even linear systematic sampling.

3.6 Circular Systematic Sampling

It has been pointed out in the beginning of this chapter, the population size is a multiple of the sample size. However in practice this requirement will not be satisfied always. Some survey practitioners will try to take the sampling interval k as the integer nearest to N/n. When this is followed, some times we may not get a sample of the desired size. For example, when N=20, n=3 and k=7, the random start 7 yields units with labels 7 and 14 as sample whereas the desired sample size is 3. In some cases, some units will never appear in the sample thereby the estimation of the population total (mean) becomes impossible. For example, when N=30, n=7 and k=4, the units with labels 29 and 30 will never appear as sampled units. These problems can be overcome by adopting a method, known as Circular Systematic Sampling (*CSS*) by Lahiri(1952). This method consists in choosing a random start from 1 to N and selecting the unit corresponding to the random start and thereafter every kth unit in a cyclical manner till a sample of size n units is obtained, k being the integer nearest to N/n. That is, if r is the number selected at random from 1 to N, the sample consists of the units corresponding to the numbers $r+jk$ if $r+jk \le N$ and

$r+jk-N$ if $r+jk>N$ for $j=0,1,...,(n-1)$ .It is to be noted that, if the sampling interval is taken as the integer closest to *N/n*, it is not always possible to get a sample of the given size as shown in the following example. Let *N*=15, *n*=6 and *k*=3. The sample corresponding to the random start 3 has only five distinct elements namely 3,6,9,12,15. Motivated by this, Kunte(1978) suggested the use of the largest integer smaller than or equal to *N/n* to avoid the above mentioned difficulty. The following theorem due to Kunte(1978) gives a necessary and sufficient condition under which one can always obtain samples having *n* distinct elements for any *n* less than or equal to *N*.

Theorem 3.9 A necessary and sufficient condition for all elements of $s(r,n)$ the sample to be distinct for all $r \le N$ and $n \le N$, is that *N* and *k* are relatively coprime , where $s(r,n)=\{i_1,i_2,....,i_n\}$. Here $i_j=[(j-1)k+r]$ mod N, with the convention that 0 is identified with N.

Proof **Sufficiency**

Suppose *N* and *k* are relatively coprime and there exists *r* and *n* such that two elements of $s(r,n)$ are equal.

Without loss of generality assume that i_1 and $i_{j+1}=[jk+r]$ mod N, where $j<n\le N$. This contradicts the fact that *k* and *N* are coprimes.

Necessity

Suppose for all $r \le N$ and $n \le N$, all elements of $s(r,n)$ are distinct and *N* and *k* are not coprimes. Let *gcd(k,N)=a*, with $k=b.a-N=c.a$, where *b* and *c* are both smaller than *N*. For any *r* let us take $n \ge c+1$. Then

$$\begin{aligned} i_{c+1} &= [ck+r] \text{ mod } N \\ &= [c.b.a+r] \text{ mod } N \\ &= [b.N+r] \text{ mod } N \\ &= r = i_1 \end{aligned}$$

This again contradicts our assumption that all elements of $s(r,n)$ are distinct. Hence the theorem. ■

Under circular systematic sampling, the conventional expansion estimator is unbiased for the population total (whenever *N* and *k* are relatively coprime) and its variance is given by $\frac{1}{N}\sum_{i=1}^{N}[\hat{Y}_{ci}-Y]^2$ where $\hat{Y}_{ci}=\frac{N}{n}\sum_{j=1}^{n}y_j^i$, y_j^i being the *y*-value of the *j*th unit in the circular systematic sampling corresponding to the random start *i*.

3.7 Systematic Sampling in Two Dimensions

The linear systematic sampling can be extended for two dimensions populations in a straightforward manner . Here it is assumed that the *nmkl* population units are arranged in the form of *ml* rows , each containing *nk* units and it is planned to select a systematic sample of mn units. The following procedure is adopted to draw a sample of size mn.

Two random numbers r and s are independently chosen from 1 to l and 1 to k respectively. Then the sample of size nm is obtained by using the units with coordinates $r+(i-1)l, s+(j-1)k, i=1,2,...,m$ and $j=1,2,...,n$. For example, when m=3, l=3, n=3 and k=4, the units corresponding to the random starts 2 and 3 are those placed against the coordinates (2,3), (5,3), (8,3), (2,7), (5,7), (8,7), (2,11), (5,11) and (8,11). Refer the Diagram 3.1 A systematic sample selected in this manner is called aligned sample.

Theorem 3.10 An unbiased estimator for the population total corresponding to the random starts r and s is given by $\hat{Y}_{TD} = \frac{NM}{nm}\sum_{i=1}^{m}\sum_{j=1}^{n} y_{r+(i-1)l, s+(j-1)k}$

Proof Note that the estimator $\hat{Y}_{TD}$ can take any one of the lk values

$$\frac{NM}{nm}\sum_{i=1}^{m}\sum_{j=1}^{n} y_{r+(i-1)l, s+(j-1)k}, r=1,2,...,l, s=1,2,...,k$$

with equal probabilities values $\frac{1}{kl}$.

Therefore $E(\hat{Y}_{TD}) = \sum_{r=1}^{l}\sum_{s=1}^{k}\frac{1}{lk}\frac{NM}{nm}\sum_{i=1}^{m}\sum_{j=1}^{n} Y_{r+(i-1)l, s+(j-1)k} = Y$

Hence the proof. ■

Remark It may be noted that the variance of the above estimator is $\frac{1}{kl}\sum_{r=1}^{l}\sum_{s=1}^{k}[\hat{Y}_{rs} - Y]^2$ where $\hat{Y}_{rs}$ is the expansion estimator defined in the above theorem corresponding to the random starts r and s.

Theorem 3.11 An aligned sample of size n drawn from a population consisting of n^2k^2 units has the same precision as a simple random sample of size n^2 when the population values are represented by the relation

$$Y_{ij} = i + j, i = 1,2,...,nk; j = 1,2,...,nk$$

if the expansion estimator is used for estimating the population total.

Proof The variance of the expansion estimator based on a sample of size n^2 drawn from a population containing n^2k^2 units is given by

$$V(\hat{Y}_{srs}) = \frac{n^4k^4(n^2k^2 - n^2)}{n^2k^2n^2}\frac{1}{n^2k^2 - 1}\sum_{i=1}^{nk}\sum_{j=1}^{nk}[Y_{ij} - \overline{Y}]^2 \quad (3.23)$$

Note that $\overline{Y} = \frac{1}{n^2k^2}\sum_{i=1}^{nk}\sum_{j=1}^{nk} Y_{ij}$

$$=\frac{1}{n^2k^2}\sum_{i=1}^{nk}\sum_{j=1}^{nk}[i+j]$$

$$=\frac{1}{n^2k^2}\left[\left\{\frac{(nk)(nk)(nk+1)}{2}\right\}+\left\{\frac{(nk)(nk)(nk+1)}{2}\right\}\right]$$

$$=nk+1$$

Therefore $[Y_{ij}-\bar{Y}]^2=[i+j-(nk+1)]^2$

Summing both sides with respect to i and j from 1 to nk we get

$$\sum_{i=1}^{nk}\sum_{j=1}^{nk}[Y_{ij}-\bar{Y}]^2=\sum_{i=1}^{nk}\sum_{j=1}^{nk}[i^2+j^2+2ij]-2(nk+1)\sum_{i=1}^{nk}\sum_{j=1}^{nk}(i+j)+n^2k^2(nk+1)^2$$

$$=2\left\{\frac{(nk)(nk+1)(2nk+1)}{6}\right\}+2\left\{\frac{(nk)(nk+1)}{2}\right\}^2-4\left\{\frac{(nk+1)(nk)(nk)(nk+1)}{2}\right\}+(n^2k^2)(nk+1)^2$$

$$=\frac{n^2k^2(n^2k^2-1)}{6} \tag{3.24}$$

Substituting (3.24) in (3.23) we get

$$V(\hat{Y}_{srs})=\frac{n^2k^2(n^2k^2-n^2)}{n^2}\frac{1}{n^2k^2-1}\frac{n^2k^2(n^2k^2-1)}{6}$$

$$=\frac{n^4k^4(n^2k^2-1)}{6} \tag{3.25}$$

Since $\hat{Y}_{TD}$ is unbiased for the population total, we have

$$V(\hat{Y}_{TD})=E(\hat{Y}_{TD}-Y)^2$$

$$=E(\hat{Y}_{TD})^2-Y^2$$

$$=E(\hat{Y}_{TD})^2-n^4k^4(nk+1)^2 \tag{3.26}$$

Corresponding to the random starts r and s we have

$$\hat{Y}_{TD}=\frac{n^2k^2}{n^2}\sum_{i=1}^{n}\sum_{j=1}^{n}Y_{r+(i-1)l,s+(j-1)k}$$

$$=k^2\sum_{i=1}^{n}\sum_{j=1}^{n}[r+(i-1)k+s+(j-1)k]$$

$$=k^2\sum_{i=1}^{n}\sum_{j=1}^{n}[r+s+(i+j)k-2k]$$

$$= k^2[n^2(r+s) + 2k\sum_{i=1}^{n}\sum_{j=1}^{n} j - 2n^2k]$$

$$= k^2n^2[r+s+k(n-1)] \tag{3.27}$$

Therefore

$$E[\hat{Y}_{TD}]^2 = \sum_{r=1}^{k}\sum_{s=1}^{k}[\hat{Y}_{TD}]^2\frac{1}{k^2}$$

$$= k^2n^4\left[2\sum_{r=1}^{k}\sum_{s=1}^{k} r^2 + 2\sum_{r=1}^{k}\sum_{s=1}^{k} rs + 4k(n-1)\sum_{r=1}^{k}\sum_{s=1}^{k} r + k^4(n-1)^2\right]$$

$$= k^2n^4\left\{\begin{matrix}\frac{2k[k(k+1)(2k+1)}{6} + 2\frac{k^2(k+1)^2}{4} + \\ 4k(n-1)\frac{k(k+1)}{2} + k^4(n-1)^2\end{matrix}\right\}$$

$$= k^4n^4\left\{\frac{(k+1)(7k+5)}{6} + k^2(n-1)(nk+k+2)\right\} \tag{3.28}$$

Substituting (3.28) in (3.26) we get and simplifying the expression we get

$$V(\hat{Y}_{TD}) = \frac{n^4k^4(n^2k^2-1)}{6}$$

Hence the proof. ■

The aligned systematic sampling described above is sometimes referred to as square grid sampling by some authors. For two dimensional populations an alternative sampling scheme known as "unaligned scheme" is also available. The details are furnished below.

Two independent sets of random integers of sizes n and m, namely $\{i_1, i_2, ..., i_n\}$ and $\{j_1, j_2, ..., j_n\}$ are drawn from 1 to l and 1 to k respectively. Then the units selected for the sample are those having the coordinates $(i_1 + rl, j_{r+1}), (i_2 + rl, j_{r+1} + k), ..., (i_n + rl, j_{r+1} + (n-1)k, r = 0, 1, ...(m-1)$ For example, the diagram 3.2 shows an unaligned sample of size nine when m=3, l=3, n=3 and k=4; in this example $i_1 = 2, i_2 = 3, i_3 = 1$, $j_1 = 2$, $j_2 = 1$ and $j_3 = 3$.

The two sampling schemes described for two dimensional populations have been compared by Quenouille (1949) and Das (1950). A review of systematic sampling in one or more dimensions is found in Bellhouse (1988).

3.8 Problems and Solutions

Problem 3.1 Derive the average variances of the expansion estimators under linear systematic sampling and simple random sampling assuming $Y_1, Y_2, ..., Y_N$ are random variables satisfying

	1	2	3	4	5	6	7	8	9	10	11	12
1												
2			♣				♣			♣		
3												
4												
5			♣				♣			♣		
6												
7												
8			♣				♣			♣		
9												

Fig. 3.1 Aligned Sample of size 9

	1	2	3	4	5	6	7	8	9	10	11	12
1										♣		
2		♣										
3						♣						
4									♣			
5	♣											
6					♣							
7											♣	
8			♣									
9								♣				

Fig. 3.2 Unaligned sample of size 9

$E_M(Y_i) = a + bi$, $V_M(Y_i) = \sigma^2$, $i = 1, 2, ..., N$ and $\text{cov}_M(Y_i, Y_j) = 0, i \neq j$.

Solution The variables Y_i 's can be written as

$$Y_i = a + bi + e_i, i = 1, 2, ..., N$$

where e_i 's satisfy

$$E_M(e_i) = 0, V_M(e_i) = \sigma^2, i = 1,2,..., N \text{ and } \text{cov}_M(e_i, e_j) = 0 \text{ for } i \neq j.$$

Under the above described model $\bar{Y} = a + \left[\frac{b(N+1)}{2}\right] + \frac{1}{N}\sum_{j=1}^{N} e_j$

Therefore $Y_i - \bar{Y} = b + \left\{i - \frac{(N+1)}{2}\right\} + \left\{e_i - \frac{1}{N}\sum_{j=1}^{N} e_j\right\}$

Squaring and summing with respect to i from 1 to N we obtain

$$E_M \sum_{i=1}^{N} [Y_i - \bar{Y}]^2 = b^2 \sum_{i=1}^{N} \left[\frac{i^2 - (N+1)i + (N+1)^2}{4}\right] + N\sigma^2 + \frac{\sigma^2}{N} - \frac{2\sigma^2}{N} \quad (3.28)$$

We know that under simple random sampling

$$V(\hat{Y}_{SRS}) = \frac{N^2(N-n)}{Nn} \frac{1}{(N-1)} \left[\sum_{i=1}^{N} [Y_i - \bar{Y}]^2\right]$$

Therefore $E_M[V(\hat{Y}_{SRS})] = \frac{N^2(N-n)}{Nn} \frac{1}{(N-1)} E_M \left[\sum_{i=1}^{N} [Y_i - \bar{Y}]^2\right]$ (3.29)

Substituting (3.28) in (3.29) and doing the necessary simplifications we get the average variance of the expansion estimator as

$$E_M[V(\hat{Y}_{SRS})] = \left[\frac{(nk+1)(k-1)}{12n}\right] b^2 + N(k-1)\sigma^2$$

We have seen in Section 3.2

$$\hat{Y}_r - Y = \frac{N}{n}\sum_{j=1}^{n} Y_{r+(j-1)k} - \sum_{i=1}^{N} Y_i$$

$$= \frac{N}{n}\sum_{j=1}^{n} \{a + b[r + (j-1)k + e_{r+(j-1)k}]\} - \sum_{i=1}^{N} \{a + bi + e_i\}$$

(substituting the model)

$$= Nb\left[\frac{r - (k+1)}{2}\right] + \frac{N}{n}\sum_{j=1}^{n} e_{r+(j-1)k} - \sum_{s=1}^{N} e_s$$

Squaring both the sides we get

$$[\hat{Y}_r - Y]^2 = N^2b^2\left[\frac{r-(k+1)}{2}\right]^2 + \frac{N^2}{n^2}\sum_{j=1}^{n} e_{r+(j-1)k}^2 -$$

$$\sum_{s=1}^{N} e_s^2 - \frac{2N}{n}\left\{\sum_{j=1}^{n} e_{r+(j-1)k}\right\}\left\{\sum_{s=1}^{N} e_s\right\}$$

Taking expectations on both the sides we get

$$E_M[\hat{Y}_r - Y]^2 = N^2b^2\left[\frac{r-(k+1)}{2}\right]^2 + \left(\frac{N^2}{n}\right)\sigma^2 + N\sigma^2 - 2N\sigma^2$$

$$= N^2b^2\left[\frac{r-(k+1)}{2}\right]^2 + N(k-1)\sigma^2 \text{ (using } N=nk) \tag{3.30}$$

We know that under systematic sampling

$$V(\hat{Y}_{LSS}) = \frac{1}{k}\sum_{r=1}^{k}[\hat{Y}_r - Y]^2 \tag{3.31}$$

Using (3.30) in (3.31) we get the average variance of the expansion estimator under linear systematic sampling as

$$E_M[V(\hat{Y}_{LSS})] = N^2\left[\frac{k^2-1}{12}\right] + N(k-1)^2$$

Hence the solution. ■

***Problem* 3.2** Let s_r denote the set of labels included in a circular systematic sample of size n and $n_1(r)$ be the number of labels in s_r for which $r+jk \le N$ and S_r be the complement set for which $r+jk \le N$. If the end corrections are applied to the sample mean $\bar{y}_r$ by giving the weight $\frac{1}{n}+x$ to the unit with smallest label, $\frac{1}{n}-x$ to the unit with largest label and $\frac{1}{n}$ to the remaining $(n-2)$ units in the sample so that the resulting estimator reduces to the population mean when there is a perfect linear trend $Y_i = a+bi$, then show that

$$x = \begin{cases} \dfrac{2r+(n-1)k-(N+1)}{2(n-1)k} \text{ if } n_1(r) = n \\ \dfrac{2r+(n-1)k-(3N+1)+2Nn_1(r)}{2n(N-k)} \text{ if } n_1(r) < n \end{cases}$$

Solution **Case (1)** $n_1(r) = n$

In this case the corrected sample mean can be written as

$$\bar{y}_c = \left[\frac{1}{n} + x\right] Y_r + \frac{1}{n} \sum_{j=1}^{n} Y_{r+(j-1)k} + \frac{1}{n} Y_{r+(n-1)k}$$

When we use the model $Y_i \doteq a + bi$, the above estimator on simplification reduces to

$$(a + br) + \left[b(n-1)\frac{k}{2}\right] + xb(n-1)k \tag{3.32}$$

We know that when $Y_i = a + bi$, the population mean is given by

$$a + b + \frac{(nk+1)}{2} \tag{3.33}$$

Comparing the coefficients of b in (3.32) and (3.33) we get

$$r + \left[(n-1)\frac{k}{2}\right] + xk(n-1) = \frac{nk+1}{2}$$

Solving for x we get $x = \dfrac{2r + (n-1)k - (N+1)}{2(n-1)k}$

Case (2) $n_1(r) < n$

In this case the n labels included in the sample can be grouped into two categories. In the first category, we have those labels for which $r + (j-1)k > N$ and in the second category, we include those labels for which $r + (j-1) \leq N$. Note that the smallest label is nothing but the first label in the group 2, namely the one which exceeds N for the first time when we make circular systematic selection and the largest label is the last label in group 1, namely the one satisfying the inequality $r + (j-1)k < N$. Therefore the corrected sample can be expressed as

$$\bar{y}_c = \frac{1}{n} \sum_{j=1}^{n_1(r)-1} Y_{r+(j-1)k} \left[\frac{1}{n} - x\right] Y_{r+[n_1(r)-1]k} + \left[\frac{1}{n} + x\right] Y_{r+n_1(r)k-N}$$

$$+ \frac{1}{n} \sum_{j=n_1(r)+2}^{n} Y_{r+(j-1)k-N}$$

$$= [Y_{r+n_1(r)k-N} - Y_{r+(n_1(r)-1)k}]x + \frac{1}{n} \sum_{j=1}^{n_1(r)} Y_{r+(j-1)k} + \sum_{j=n_1(r)+1}^{n} Y_{r+(j-1)k} \tag{3.34}$$

Substituting the model values we get

$$Y_{r+n_1(r)k-N} - Y_{r+(n_1(r)-1)k} = bk(1-n) \tag{3.35}$$

Note that

$$\sum_{j=1}^{n_1(r)} Y_{r+(j-1)k} + \sum_{j=n_1(r)+1}^{n} Y_{r+(j-1)k-N}$$

$$=\frac{1}{n}\left[\sum_{j=1}^{n}Y_{r+(j-1)k}-bN[n-n_1(r)]\right]$$

$$=a+br+b(n-1)\frac{k}{2}-b\frac{N}{n}[n-n_1(r)] \tag{3.36}$$

Using (3.35) and (3.36) in (3.34) we get

$$\bar{y}_c=bk(1-n)x+a+br+b(n-1)\frac{k}{2}-bN\frac{n-n_1(r)}{n} \tag{3.37}$$

Comparing the coefficient of b in (3.37) with the same in the population mean we get

$$\frac{nk+1}{2}=k(1-n)x+r+(n-1)\frac{k}{2}-[n-n_1(r)]\frac{N}{n}$$

Solving the above equation for x we get the required expression. ■

Problem 3.3 Show that the sample mean coincides with the population mean in the presence of linear trend when a purposive sample of size 2 in which the first and last population units are included. Compare its average mean and square error of the sample mean in centered systematic sampling assuming k is odd with the help of the super-population model

$$Y_i=a+bi+e_i\,,i=1,2,...,N$$

where e_i's satisfy $E_M(e_i)=0$, $E_M(e_i^2)=\sigma^2$ and $E_M(e_ie_j)=0,i\neq j$

Solution Note that, if $Y_i=a+bi+,i=1,2,...,N$, the sample mean above mentioned purposive sampling scheme is

$$\bar{y}_p=\frac{Y_1+Y_N}{2}$$

$$=\frac{a+b+a+bN}{2}$$

$$=a+b\frac{N+1}{2}$$

The right hand side of the above expression is nothing but the population mean in the presence of linear trend. Under the super-population model described in the problem we have

(1) $\bar{y}_p=a+b\left[\frac{N+1}{2}\right]+\frac{1}{2}(e_1+e_N)$

(2) $\bar{y}_c=a+b\left[\frac{N+1}{2}\right]+\frac{1}{n}\sum_{j=1}^{n}e_{\frac{k+1}{2}+(j-1)k}$ and

(3) $\bar{Y}=a+b\left[\frac{N+1}{2}\right]+\frac{1}{N}\sum_{j=1}^{N}e_j$

where $\bar{y}_c$ is the sample mean under centered systematic sampling. Since both the sampling scheme considered above are purposive sampling schemes, we have

$$E_M[MSE(\bar{y}_p)] = E_M[(\bar{y}_p - \bar{Y})^2] = E_M\left[\frac{1}{2}(e_1 + e_N) - \frac{1}{N}\sum_{j=1}^{N} e_j\right]^2$$

$$= E_M\left[\frac{1}{4}(e_1^2 + e_N^2 + 2e_1 e_N) + \frac{1}{N^2}\sum_{j=1}^{N} e_j^2\right] +$$

$$E_M\left[\frac{2}{N^2}\sum_{i<j} e_i e_i - \frac{1}{N}(e_1 + e_N)\sum_{j=1}^{N} e_j\right]$$

$$= \frac{1}{4}(2\sigma^2) + \frac{1}{N^2}N\sigma^2 - \frac{1}{N}(2\sigma^2)$$

$$= \frac{N-2}{2N}\sigma^2$$

$$E_M[MSE(\bar{y}_c)] = E_M[\ \bar{y}_c - \bar{Y}\ ]^2$$

$$= E_M\left[\left(\frac{1}{n}\right)\sum_{j=1}^{n} e_{\frac{(k+1)}{2}+(j-1)k} - \left(\frac{1}{N}\right)\sum_{j=1}^{N} e_j\right]^2$$

$$= \left[\frac{1}{n}\right]^2 E_M \sum_{j=1}^{n} e_{\frac{(k+1)}{2}+(j-1)k} + \left[\frac{1}{N}\right]^2 E_M \sum_{j=1}^{n} e_j^2 - \left[\frac{2}{Nn}\right]$$

$$E_M\left\{\sum_{j=1}^{n} e_{\frac{(k+1)}{2}+(j-1)k}\right\}\left\{\sum_{j=1}^{N} e_j\right\}$$

$$= \left[\frac{1}{n}\right]^2 n\sigma^2 + \left[\frac{1}{N}\right]^2 n\sigma^2 - \left[\frac{2}{Nn}\right] n\sigma^2$$

$$= \frac{N-n}{Nn}\sigma^2$$

It can be seen that $E_M[\ MSE(\bar{y}_p)\ ] - E_M[MSE(\bar{y}_c)\ ] \geq 0$. Hence we infer that among these two purposive sampling schemes, centered systematic sampling is more efficient than the two point scheme. ■

Exercises

3.1 Derive average mean squared errors under balanced systematic and modified systematic sampling schemes under the super-population model described in Problem 3.4 and compare them.

3.2 Derive the variance of the conventional estimators in simple random sampling, Linear systematic sampling and stratified sampling assuming $Y_i = a + bi + ci^2$ and compare them.

3.3 Develop corrected estimator for the population mean in the presence of parabolic trend assuming the sample size is odd under Linear systematic sampling.(Hint: The first, middle most and last units in the sample can be corrected so that the corrected estimator coincides with population mean.)

3.4 Assuming the population values are independently distributed with same mean and same variance, show that the mean square error of the estimate based on a systematic sample is increased on the average by applying the end corrections to the extent $\frac{k^2(k^2-1)}{6(n-1)^2}\sigma^2$.

3.5 Derive the variance of conventional estimator for the population mean under circular systematic sampling assuming the presence of linear trend.

Chapter 4

Unequal Probability Sampling

4.1 PPSWR Sampling Method

In simple random sampling and systematic sampling selection probabilities are equal for all units in the population. These schemes do not take into account the inherent variation in the values of population units. Therefore, these methods are likely to yield results which are not totally reliable, particularly when the units have significant variation with respect to their values. For such populations, one can think of other sampling schemes, provided additional information about a suitable variable is available for all the units in the population. These type of variables are called 'size' variables. In this section, we shall present a method which makes use of the size information.

Let X_i and Y_i be the values of the size variable and the study variable for the ith population unit, $i = 1, 2, \ldots, N$. Here it is assumed that all the N values $X_1, X_2, \ldots, X_N$ are known. A sample of size n is obtained in the form of n units with replacement draws, where in each draw the probability of selecting the unit i, namely P_i is proportional to its size X_i. A sample obtained in this manner is known as Probability Proportional to Size With Replacement ($PPSWR$) sample.

It is to be noted that $P_i \propto X_i \Rightarrow P_i = kX_i$. Summing both sides we get $\sum_{i=1}^{N} P_i = k\sum_{i=1}^{N} X_i$. Therefore $k = \frac{1}{X}$ and hence $P_i = \frac{X_i}{X}$ for $i = 1, 2, \ldots, N$.

In order to select units with these probabilities, one can use either "Cumulative Total Method" or "Lahiri's Method". These methods are described below.

Cumulative Total Method

Obtain the cumulative totals

$$T_1 = X_1,\ T_2 = X_1 + X_2,\ T_3 = X_1 + X_2 + X_3$$

$$\ldots \quad \ldots$$

$$T_N = X_1 + X_2 + \ldots + X_N$$

A random number R is drawn from 1 to N. If $T_{i-1} < R \le T_i$, then the ith unit is selected. It can be seen that in this method, the probability of selecting the ith unit in a given draw is $\frac{T_i - T_{i-1}}{X}$ which is nothing but $\frac{X_i}{X}$. This procedure is repeated n times to get a sample of size n.

Note Sometimes X_i's can take even non-integral values. In such cases a new set of values can be derived by multiplying each X_i by a suitable power of 10 and the resulting values can be used in the selection process.

Lahiri's Method

The cumulative total method described above is very difficult to implement for populations having larger number of units. In such cases one can employ the Lahiri's method of *PPS* selection which is described below.

Let M be an integer greater than or equal to the maximum of the sizes $X_1, X_2, \ldots, X_N$. The following two steps are executed to select one unit.

Step 1 Draw a random number say i from 1 to N.

Step 2 Draw a random number R from 1 to M.

If $R \le X_i$, then the ith unit is selected. If $R > X_i$, then steps 1 and 2 are repeated till we get one unit.

Theorem 4.1 The probability of selecting the ith unit in the first effective draw is $\frac{X_i}{X}$, in Lahiri's method of *PPS* selection.

Proof Note that a draw becomes ineffective when the number drawn in step 1 is i and that drawn in step 2 exceeds X_i. Therefore the probability of a draw becoming ineffective is

$$\lambda = \sum_{i=1}^{N} \frac{1}{N}\left[\frac{M - X_i}{M}\right]$$

$$= 1 - \frac{\overline{X}}{M}$$

This implies the probability of selecting the ith unit in the first effective draw is

$$\left[\frac{X_i}{NM}\right] + \left[\frac{X_i}{NM}\right]\lambda + \left[\frac{X_i}{NM}\right]\lambda^2 + \ldots$$

$$= \left[\frac{X_i}{NM}\right][1-\lambda]^{-1} = \left[\frac{X_i}{NM}\right]\left[1 - 1 + \frac{\overline{X}}{M}\right]^{-1} = \left[\frac{X_i}{X}\right]$$

Hence the proof. ■

Thus from the above theorem we infer that Lahiri's method of *PPS* selection yields desired selection probabilities. The following theorem gives an unbiased estimator for the population total under *PPSWR* and also its variance estimator.

***Theorem* 4.2** Let y_i be the y-value of the unit drawn in the ith draw and p_i be the corresponding selection probability, $i = 1, 2, \dots, n$. Then an unbiased estimator for population total is $\hat{Y}_{pps} = \frac{1}{n}\sum_{i=1}^{n}\frac{y_i}{p_i}$ (4.1)

and its variance is $V[\hat{Y}_{pps}] = \frac{1}{n}\sum_{i=1}^{N}\left\{\frac{Y_i}{P_i} - Y\right\}^2 P_i$ (4.2)

Proof Since the ratio $\frac{y_i}{p_i}$ can take any one of the N values $\frac{Y_j}{P_j}$, $j = 1, 2, \dots, N$ with respective probabilities P_i,

$$E\left[\frac{y_i}{p_i}\right] = \sum_{j=1}^{N}\left[\frac{Y_j}{P_j}\right]P_j = Y$$

Hence $E[\hat{Y}_{pps}] = E\left[\frac{1}{n}\sum_{i=1}^{n}\frac{y_i}{p_i}\right] = \left[\frac{1}{n}\sum_{i=1}^{n}E\left[\frac{y_i}{p_i}\right]\right]$

$$= \left[\left(\frac{1}{n}\right)nY\right] = Y$$

Therefore $\hat{Y}_{pps} = \frac{1}{n}\sum_{i=1}^{n}\frac{y_i}{p_i}$ is unbiased for the population total under *PPSWR*.

Since the draws are independent,

$$V[\hat{Y}_{pps}] = V\left[\frac{1}{n}\sum_{i=1}^{n}\frac{y_i}{p_i}\right]$$

$$= \frac{1}{n^2}\sum_{i=1}^{n}V\left[\frac{y_i}{p_i}\right]$$

$$= \left[\frac{1}{n^2}\sum_{i=1}^{n}E\left[\frac{y_i}{p_i} - Y\right]^2\right] \tag{4.3}$$

Note that, the quantity $\left[\frac{y_i}{p_i} - Y\right]^2$ can take any one of the N values $\left[\frac{Y_j}{P_j} - Y\right]^2$ with respective probabilities P_j, $j = 1, 2, \dots, N$

Therefore $E\left[\frac{y_i}{p_i} - Y\right]^2 = \sum_{j=1}^{N}\left[\frac{Y_j}{P_j} - Y\right]^2 P_j$ (4.4)

Using (4.4) in (4.3) we get $V[\hat{Y}_{pps}] = \frac{1}{n}\sum_{i=1}^{N}\left\{\frac{Y_i}{P_i} - Y\right\}^2 P_i$

Hence the proof. ■

The following theorem gives an unbiased estimator of $V[\hat{Y}_{pps}]$.

Theorem 4.3 An unbiased estimator of $V[\hat{Y}_{pps}]$ is

$$v[\hat{Y}_{pps}] = \frac{1}{n(n-1)}\sum_{i=1}^{n}\left\{\frac{y_i}{p_i} - \hat{Y}\right\}^2 p_i$$

Proof By Theorem 4.2,

$$V[\hat{Y}_{pps}] = \frac{1}{n}\sum_{i=1}^{N}\left\{\frac{Y_i}{P_i} - Y\right\}^2 P_i$$

$$= \frac{1}{n}\sum_{i=1}^{N}\frac{Y_i^2}{P_i} - Y^2 \tag{4.5}$$

If $v[\hat{Y}_{pps}]$ is an unbiased estimator of $V[\hat{Y}_{pps}]$ then

$$E\{\, v[\hat{Y}_{pps}]\,\} = V[\hat{Y}_{pps}] \tag{4.6}$$

Since $V[\hat{Y}_{pps}] = E[\hat{Y}_{pps}^2] - Y^2$, we have

$$E\{\, v[\hat{Y}_{pps}]\,\} = E[\hat{Y}_{pps}^2] - Y^2 \quad \text{(using (4.6))}$$

Hence $$Y^2 = E[\hat{Y}_{pps}^2 - v(\hat{Y}_{pps})] \tag{4.7}$$

Note that $$E\left\{\frac{1}{n}\sum_{i=1}^{n}\frac{y_i^2}{p_i}\right\} = \sum_{i=1}^{N}\frac{Y_i^2}{P_i} \tag{4.8}$$

Using (4.6),(4.7) and (4.8) in (4.5) we get

$$E\{\, v[\hat{Y}_{pps}]\,\} = \frac{1}{n}E\left\{\frac{1}{n}\sum_{i=1}^{n}\frac{y_i^2}{p_i}\right\} - \frac{1}{n}E[\hat{Y}_{pps}^2 - v(\hat{Y}_{pps})]$$

Therefore $$\frac{n-1}{n}E\{v[\hat{Y}_{pps}]\} = E\left\{\frac{1}{n^2}\sum_{i=1}^{n}\frac{y_i^2}{p_i^2} - \frac{1}{n}\hat{Y}_{pps}^2\right\}$$

$$E\{\, v[\hat{Y}_{pps}]\,\} = E\left\{\frac{1}{(n-1)}\left[\frac{1}{n}\sum_{i=1}^{n}\frac{y_i^2}{p_i^2} - \frac{1}{n}\hat{Y}_{pps}^2\right]\right\}$$

$$= E\left\{\frac{1}{n(n-1)}\left[\sum_{i=1}^{n}\left(\frac{y_i}{p_i} - \hat{Y}_{pps}\right)^2\right]\right\}$$

Hence $\frac{1}{n(n-1)}\sum_{i=1}^{n}\left(\frac{y_i}{p_i}-\hat{Y}_{pps}\right)^2$ is an unbiased estimator of $V[\hat{Y}_{pps}]$. ∎

A meaningful equal probability sampling competitor to the *PPSWR* sampling scheme is simple random sampling with replacement. The following theorem gives an estimate for the gain due to *PPSWR* when compared to simple random sampling with replacement.

Theorem 4.4 An unbiased estimator of the gain due to *PPSWR* sampling as compared to *SRSWR* is $\frac{1}{n^2}\sum_{i=1}^{n}\left[N-\frac{1}{p_i}\right]\frac{y_i^2}{p_i}$

Proof We know that under *SRSWR*,

$$V[\hat{Y}_{swr}] = \frac{N^2(N-1)}{Nn}\frac{1}{N-1}\sum_{i=1}^{N}[Y_i-\overline{Y}]^2$$

$$= \frac{1}{n}\left\{N\sum_{i=1}^{N}Y_i^2 - Y^2\right\} \tag{4.9}$$

Note that, under *PPSWR*, unbiased estimators of the quantities $\sum_{i=1}^{N}Y_i^2$ and Y^2 are $\frac{1}{n}\sum_{i=1}^{n}\frac{y_i^2}{p_i}$ and $E[\hat{Y}_{pps}^2 - v(\hat{Y}_{pps})]$ respectively.

Therefore , by (4.9), unbiased estimator of $V[\hat{Y}_{swr}]$ under *PPSWR* is

$$\frac{1}{n^2}\left\{N\sum_{i=1}^{n}\frac{y_i^2}{p_i} - n\hat{Y}_{pps}^2\right\} + \frac{1}{n}v(\hat{Y}_{pps}) \tag{4.10}$$

Already we have seen in Theorem 4.1, an unbiased estimator of $V[\hat{Y}_{pps}]$ is

$$\frac{1}{n(n-1)}\left[\sum_{i=1}^{n}\frac{y_i^2}{p_i^2} - \hat{Y}_{pps}^2\right] \tag{4.11}$$

Subtracting (4.11) from (4.10), one can estimate the gain due to *PPSWR* as

$$\frac{1}{n^2}\sum_{i=1}^{n}\left[N-\frac{1}{p_i}\right]\frac{y_i^2}{p_i}$$

Hence the proof. ∎

4.2 PPSWOR Sampling Method

When probability proportional to size selection is made in each draws without replacing the units drawn, we get a Probability Proportional to Size Sample Without Replacement *(PPSWOR)*. Since the selection probabilities changes from draw to draw, we must device suitable estimators taking into account this aspect. In this section, we shall discuss three estimators suitable for *PPSWOR*.

Desraj Ordered estimator

Let

$$t_1 = \frac{y_1}{p_1}, t_2 = y_1 + \frac{y_2}{p_2}(1-p_1), t_3 = y_1 + y_2 + \frac{y_3}{p_3}(1-p_1-p_2), \ldots,$$

$$t_n = y_1 + y_2 + \ldots + y_{n-1} + \frac{y_n}{p_n}(1-p_1-\ldots-p_{n-1})$$

where y_i and p_i, $i = 1, 2, \ldots, n$ are as defined in Section 4.1.The Desraj ordered estimator for the population total is defined as

$$\hat{Y}_{DR} = \frac{1}{n}\sum_{i=1}^{n} t_i$$

Theorem 4.5 Under *PPSWOR*, $\hat{Y}_{DR} = \frac{1}{n}\sum_{i=1}^{n} t_i$ is unbiased for the population total and an unbiased estimator of its variance is

$$v(\hat{Y}_{DR}) = \frac{1}{n(n-1)}\sum_{i=1}^{n}(t_i - \bar{t})^2 \quad \text{where } \bar{t} = \frac{1}{n}\sum_{i=1}^{n} t_i$$

Proof Note that the ratio $\frac{y_1}{p_1}$ can take any one of the *N* values $\frac{Y_j}{P_j}$, $j=1,2,\ldots, N$ with respective probabilities P_j. Therefore

$$E\left[\frac{y_r}{p_r}\right] = \sum_{r=1}^{N}\frac{Y_r}{P_r}P_r = Y$$

Hence t_1 is unbiased for the population total.

For $r = 1, 2, \ldots, n$, $E[t_r] = E_1 E_2[t_r \mid i_1, i_2, \ldots, i_{r-1}]$

where E_2 is the conditional expectation after fixing the units with labels $i_1, i_2, \ldots, i_{r-1}$ for the first $(r-1)$ draws. Now $E[t_r]$ can be written as

$$E[t_r] = E_1\left[Y_{i_1} + Y_{i_2} + \ldots + Y_{i_{r-1}} + (1 - P_{i_1} - P_{i_2} - \ldots - P_{i_{r-1}}) E\left\{\frac{y_r}{p_r} \mid i_1, i_2, \ldots, i_{r-1}\right\}\right] \tag{4.12}$$

Note that conditionally, the ratio $\frac{y_r}{p_r}$ can take any one of the $N-r-1$ values $\frac{Y_j}{P_j}$, $j=1,2,\ldots, N, \neq i_1, i_2, \ldots, i_{r-1}$ with probabilities $\frac{P_j}{(1-P_{i_1}-P_{i_2}-\ldots-P_{i_{r-1}})}$.

Therefore $E\left\{\frac{y_r}{p_r} \mid i_1, i_2, \ldots, i_{r-1}\right\} = \sum_{\substack{j=1 \\ \neq i_1, i_2, \ldots, i_{r-1}}}^{N} \frac{Y_j}{P_j} \frac{P_j}{(1-P_{i_1}-P_{i_2}-\ldots-P_{i_{r-1}})}$

(4.13)

Substituting (4.13) in (4.12) we get

$$E(t_r) = E_1\left[Y_{i_1} + Y_{i_2} + \ldots + Y_{i_{r-1}} + \sum_{\substack{j=1 \\ \neq i_1, i_2, \ldots, i_{r-1}}}^{N} Y_j\right]$$

$$= Y$$

Thus we have, for $r = 1, 2, \ldots, n$, $E[t_r] = Y$.

Therefore $E(\hat{Y}_{DR}) = \frac{1}{n}\sum_{i=1}^{n} E(t_i)$

$$= \frac{1}{n} nY = Y$$

Hence $\hat{Y}_{DR} = \frac{1}{n}\sum_{i=1}^{n} t_i$ is unbiased for the population total.

We know that $V(\hat{Y}_{DR}) = E(\hat{Y}_{DR}^2) - Y^2$ (4.14)

Note that

$E[t_r t_s] = E_1 E_2[t_r t_s \mid i_1, i_2, \ldots, i_{s-1}]$ (assuming without any loss of generality $r<s$)

$$= E_1[t_r E_2(t_s \mid i_1, i_2, \ldots, i_{s-1})]$$

$$= E_1[t_r Y]$$

$$= Y E_1[t_r]$$

$$= Y^2$$

$$E\left\{\frac{1}{n(n-1)}\sum_{r \neq s}^{n} t_r t_s\right\} = Y^2 \quad (4.15)$$

Substituting (4.15) in (4.14) we get

$$V(\hat{Y}_{DR}) = E\left\{\frac{1}{n}\sum_{i=1}^{n} t_i\right\}^2 - E\left\{\frac{1}{n(n-1)}\sum_{r \neq s}^{n} t_r t_s\right\} - Y^2$$

$$= E\left\{\frac{1}{n^2}\sum_{r=1}^{n} t_r^2 + \left[\frac{1}{n^2} - \frac{1}{n(n-1)}\right]\sum_{r\neq s}^{n} t_r t_s\right\}$$

$$= E\left\{\frac{1}{n^2}\sum_{r=1}^{n} t_r^2 - \frac{1}{n^2(n-1)}\left[\left(\sum_{r=1}^{n} t_r\right)^2 - \sum_{r=1}^{n} t_r^2\right]\right\}$$

$$= E\left\{\frac{1}{n(n-1)}\sum_{r=1}^{n} t_r^2 - \frac{1}{n-1}\bar{t}^2\right\} \text{ where } \bar{t} = \frac{1}{n}\sum_{i=1}^{n} t_i$$

$$= E\left\{\frac{1}{n(n-1)}\sum_{r=1}^{n} (t_r - \bar{t})^2\right\}$$

Hence the proof. ∎

Murthy's Ordered Estimator

The Desraj ordered estimator depends upon the order in which the units are drawn. Murthy (1957) obtained the unordered estimator corresponding to Desraj ordered estimator. For the sake of simplicity, we shall restrict to samples of size 2 only. Suppose y_1 and y_2 are the values of the units selected in the first and second draws and p_1 and p_2 the corresponding initial selection probabilities. The ordered estimator is

$$\hat{Y}_{DR}^{(1,2)} = \frac{1}{2}\left[\frac{y_1}{p_1} + y_1 + \frac{y_2}{p_2}(1-p_1)\right]$$

$$= \frac{1}{2}\left[\frac{1+p_1}{p_1} y_1 + \frac{y_2}{p_2}(1-p_1)\right]$$

On the other hand, if the same two units are in the other order then the corresponding ordered estimator is

$$\hat{Y}_{DR}^{(2,1)} = \frac{1}{2}\left[\frac{1+p_2}{p_2} y_2 + \frac{y_1}{p_1}(1-p_2)\right]$$

Their corresponding selection probabilities are $P(1,2) = \dfrac{p_1 p_2}{(1-p_1)}$ and $P(2,1) = \dfrac{p_1 p_2}{(1-p_2)}$. The ordered estimator based on the ordered estimators $\hat{Y}_{DR}^{(1,2)}$ and $\hat{Y}_{DR}^{(2,1)}$ is given by

$$\hat{Y}_M = \frac{\hat{Y}_{DR}^{(1,2)} + \hat{Y}_{DR}^{(2,1)}}{P(1,2) + P(2,1)}$$

$$= \frac{(1-p_2)\dfrac{y_1}{p_1} + (1-p_1)\dfrac{y_2}{p_2}}{2 - p_1 - p_2}$$

An unbiased estimator of $V(\hat{Y}_M)$ is

$$\frac{(1-p_1-p_2)(1-p_1)(1-p_2)}{(2-p_1-p_2)^2}\left[\frac{y_1}{p_1}-\frac{y_2}{p_2}\right]^2$$

Horvitz-Thompson Estimator

To estimate the population total one can use the Horvitz-Thompson estimator provided the inclusion probabilities are available. In the case of *PPSWOR* explicit expressions are not available for inclusion probabilities. With the help of computers one can list all possible outcomes when n draws are made and hence calculate the inclusion probabilities. In the following sections some unequal probability sampling schemes yielding samples of distinct units are presented.

4.3 Random group method

The random group method is due to Rao,Hartley and Cochran (1962). This method makes use of the size information and always yields sample containing distinct units. In this method, the population is randomly divided into n mutually exclusive and exhaustive groups of sizes $N_1, N_2, ..., N_n$ and one unit is drawn from each group with probabilities proportional to size of the units in that group. Here the group sizes $N_1, N_2, ..., N_n$ are predetermined constants.

An unbiased estimator of the population total is

$$\hat{Y}_{RHC} = \sum_{i=1}^{n} \frac{y_i}{p_i'}$$

where y_i is the y-value of the unit drawn from the ith random group and p_i' is the selection probability of the unit drawn from the ith random group.

Let Y_{ij} and X_{ij} be the y and x value of the jth unit in the ith random group for a given partition. Then y_i can take any one of N_i values $Y_{ij}, j = 1,2,...,N_i$ and p_i' can take any one of the N_i values $\dfrac{X_{ij}}{\sum_{j=1}^{N_i} X_{ij}}, j = 1, 2, ..., N_i; i = 1, 2, ..., n$

Theorem 4.6 The estimator $\hat{Y}_{RHC} = \sum_{i=1}^{n} \frac{y_i}{p_i'}$ is unbiased for population total Y.

Proof $E[\hat{Y}_{RHC}] = E_1 E_2[\hat{Y}_{RHC} \mid G_1, G_2, ..., G_n]$ (4.16)

where E_2 is the conditional expectation taken with respect to a given partitioning of the population and E_1 is the overall expectation.

Note that $E_2[\hat{Y}_{RHC} \mid G_1, G_2, ..., G_n] = E_2\left\{\sum_{i=1}^{n} \frac{y_i}{p'_i} \mid G_1, G_2, ..., G_n\right\}$

$$= \sum_{i=1}^{n} E_2\left(\frac{y_i}{p'_i} \mid G_i\right) \tag{4.17}$$

Since the ratio $\frac{y_i}{p'_i}$ can take any one of the N_i values $\dfrac{\frac{Y_{ij}}{X_{ij}}}{\sum_{j=1}^{N_i} X_{ij}}$, $j = 1,2,..., N_i$

with respective probabilities $\dfrac{X_{ij}}{\sum_{j=1}^{N_i} X_{ij}}$, we have

$$E_2\left(\frac{y_i}{p'_i} \mid G_i\right) = \sum_{j=1}^{N_i} \frac{Y_{ij}}{\frac{X_{ij}}{\sum_{j=1}^{N_i} X_{ij}}} \frac{X_{ij}}{\sum_{j=1}^{N_i} X_{ij}} = \sum_{j=1}^{N_i} Y_{ij} \tag{4.18}$$

Substituting (4.18) in (4.17) we get

$$E_2[\hat{Y}_{RHC} \mid G_1, G_2, ..., G_n] = \sum_{i=1}^{n} \sum_{j=1}^{N_i} Y_{ij}$$
$$= Y$$

Therefore, by (4.16), $\hat{Y}_{RHC} = \sum_{i=1}^{n} \frac{y_i}{p'_i}$ is unbiased for the population total under random group method. ■

The following theorem gives the variance of the estimator $\hat{Y}_{RHC}$

Theorem 4.7 The variance of the estimator $\hat{Y}_{RHC}$ is

$$\frac{\sum_{i=1}^{n} N_i(N_i - 1)}{N(N-1)} \sum_{r=1}^{N} \left[\frac{Y_r}{P_r} - Y\right]^2 P_r$$

Proof

$$V[\hat{Y}_{RHC}] = E_1 V_2[\hat{Y}_{RHC} \mid G_1, G_2, ..., G_n] + V_1 E_2[\hat{Y}_{RHC} \mid G_1, G_2, ..., G_n] \tag{4.19}$$

We have seen in Theorem 4.6, $E_2[\hat{Y}_{RHC} \mid G_1, G_2, ..., G_n] = Y$

Therefore $V_1 E_2[\hat{Y}_{RHC} \mid G_1, G_2, ..., G_n] = 0$ (4.20)

Since draws are made independently in different groups ,

$$V_2[\hat{Y}_{RHC} \mid G_1, G_2, \ldots, G_n] = V_2\left\{\sum_{i=1}^{n} \frac{y_i}{p_i'} \mid G_1, G_2, \ldots, G_n\right\}$$

$$= \sum_{i=1}^{n} V_2\left(\frac{y_i}{p_i'} \mid G_i\right)$$

$$= \sum_{i=1}^{n} \sum_{j=1}^{N_i} \left[\frac{Y_{ij}}{P_{ij}} - T_i\right]^2 P_{ij} \tag{4.21}$$

where $P_{ij} = \frac{X_{ij}}{\sum_{k=1}^{N_i} X_{ik}}$ and $T_i = \sum_{k=1}^{N_i} X_{ik}$. The right hand side of the above expression is obtained by applying Theorem 4.1 with n=1.

Claim $\sum_{i=1}^{N}\left[\frac{Y_i}{P_i} - Y\right]^2 P_i = \sum_{i<j}^{N}\left[\frac{Y_i}{P_i} - \frac{Y_j}{P_j}\right]^2 P_i P_j$

Proof of the claim

We know that $\sum_{i=1}^{n}\sum_{j=1}^{n} a_{ij} = \sum_{i=1}^{n} a_{ii} + 2\sum_{i<j}^{n} a_{ij}$ if $a_{ij} = a_{ji}$

Therefore

$$\sum_{i=1}^{N}\sum_{j=1}^{N}\left[\frac{Y_i}{P_i} - \frac{Y_j}{P_j}\right]^2 P_i P_j = \sum_{i=1}^{N}\left[\frac{Y_i}{P_i} - \frac{Y_j}{P_i}\right]^2 P_i P_i + 2\sum_{i<j}^{N}\left[\frac{Y_i}{P_i} - \frac{Y_j}{P_j}\right]^2 P_i P_j \tag{4.22}$$

The above expression can be written as

$$2\sum_{i=1}^{N}\sum_{j=1}^{N}\frac{Y_i^2}{P_i^2} P_i P_j - 2\sum_{i=1}^{N}\sum_{j=1}^{N} Y_i Y_j = 2\sum_{i<j}^{N}\left[\frac{Y_i}{P_i} - \frac{Y_j}{P_j}\right]^2 P_i P_j$$

Therefore $2\sum_{i=1}^{N}\frac{Y_i^2}{P_i} - 2Y^2 = 2\sum_{i<j}^{N}\left[\frac{Y_i}{P_i} - \frac{Y_j}{P_j}\right]^2 P_i P_j$

Hence $\sum_{i=1}^{N}\left[\frac{Y_i}{P_i} - Y\right]^2 P_i = \sum_{i<j}^{N}\left[\frac{Y_i}{P_i} - \frac{Y_j}{P_j}\right]^2 P_i P_j$

Thus we have proved the claim.
Making use of (4.22) in (4.21) we get

$$V_2[\hat{Y}_{RHC} \mid G_1, G_2, \ldots, G_n] = \sum_{i=1}^{n}\sum_{j<k}^{N_i}\left[\frac{Y_{ij}}{P_{ij}} - \frac{Y_{ik}}{P_{ik}}\right]^2 P_{ij} P_{ik}$$

Since $\left[\frac{Y_{ij}}{P_{ij}}-\frac{Y_{ik}}{P_{ik}}\right]^2 P_{ij}P_{ik}$ can take any one of the values

$\left[\frac{Y_r}{P_r}-\frac{Y_s}{P_s}\right]^2 P_rP_s, r,s=1,2,...,N; r<s$ with equal probabilities $\frac{2}{N(N-1)}$,

Therefore

$$E_1V_2[\hat{Y}_{RHC} \mid G_1,G_2,...,G_n]=\sum_{i=1}^{n}\sum_{j<k}^{N_i}\frac{2}{N(N-1)}\sum_{r<s}^{N}\left[\frac{Y_r}{P_r}-\frac{Y_s}{P_s}\right]^2 P_rP_s$$

$$=\sum_{i=1}^{n}\sum_{j<k}^{N_i}\frac{2}{N(N-1)}\sum_{r=1}^{N}\left[\frac{Y_r}{P_r}-Y\right]^2 P_r$$

$$=\sum_{i=1}^{n}\frac{N_i(N_i-1)}{2}\frac{2}{N(N-1)}\sum_{r=1}^{N}\left[\frac{Y_r}{P_r}-Y\right]^2 P_r$$

$$=\sum_{i=1}^{n}\frac{N_i(N_i-1)}{N(N-1)}\sum_{r=1}^{N}\left[\frac{Y_r}{P_r}-Y\right]^2 P_r \tag{4.24}$$

Substituting (4.20) and (4.24) in (4.19) we get the required result. ∎

Remark When the groups are all of the same size, then

$$N_1=N_2=.....=N_n=\frac{N}{n}$$

In such cases we have $\sum_{i=1}^{n}N_i(N_i-1)=\frac{N(N-n)}{n}$

Substituting this in (4.24) , we obtain

$$V(\hat{Y}_{RHC})=\frac{N(N-n)}{nN(N-1)}\sum_{r=1}^{N}\left[\frac{Y_r}{P_r}-Y\right]^2 P_r$$

$$=\frac{(N-n)}{(N-1)}V(\hat{Y}_{PPS}) \text{ (refer Theorem 4.2)}$$

From this we infer that random group method is better than probability proportional to size with replacement whenever the groups are of the same size. The following theorem gives an unbiased estimator of $V(\hat{Y}_{RHC})$.

Theorem 4.8 An unbiased estimator of $V(\hat{Y}_{RHC})$ is

$$v(\hat{Y}_{RHC}) = \frac{\sum_{i=1}^{n} N_i^2 - N}{N^2 - \sum_{i=1}^{n} N_i^2}\left\{\sum_{i=1}^{n} \frac{y_i^2}{p_i p'_i} - \hat{Y}_{RHC}\right\}$$

where p'_i and p_i are as defined in Theorem 4.6 and Theorem 4.2.

Proof From Theorem 4.7 we have

$$V(\hat{Y}_{RHC}) = \lambda \sum_{r=1}^{N}\left[\frac{Y_r}{P_r} - Y\right]^2 P_r \text{ where } \lambda = \sum_{i=1}^{n} \frac{N_i(N_i - 1)}{N(N-1)}$$

$$= \lambda\left\{\sum_{r=1}^{N} \frac{Y_r^2}{P_r} - Y^2\right\} \tag{4.25}$$

Using the argument given in Theorem 4.6 it can be seen that

$$E\left\{\sum_{i=1}^{n} \frac{y_i^2}{p_i p'}\right\} = \sum_{r=1}^{N} \frac{Y_r^2}{P_r} \tag{4.26}$$

If $v[\hat{Y}_{RHC}]$ is an unbiased estimator of for $V[\hat{Y}_{RHC}]$ then

$$E\{\ v[\hat{Y}_{RHC}]\ \} = V[\hat{Y}_{RHC}] \tag{4.27}$$

Further $V[\hat{Y}_{RHC}] = E[\hat{Y}_{RHC}^2] - Y^2$

Hence $Y^2 = E[\hat{Y}_{RHC}^2 - v(\hat{Y}_{RHC})]$ (4.28)

Using (4.26),(4.27) and (4.28) in (4.25) we get

$$E\{\ v[\hat{Y}_{RHC}]\ \} = \lambda E\left\{\sum_{i=1}^{n} \frac{y_i^2}{p_i p'} + v(\hat{Y}_{RHC}) - \hat{Y}_{RHC}^2\right\}$$

Solving for $v(\hat{Y}_{RHC})$, we get as estimator of $V(\hat{Y}_{RHC})$,

$$v(\hat{Y}_{RHC}) = \frac{\lambda}{1-\lambda}\left\{\sum_{i=1}^{n} \frac{y_i^2}{p_i p'} - \hat{Y}_{RHC}^2\right\}$$

$$= \frac{\sum_{i=1}^{n} N_i^2 - N}{N^2 - \sum_{i=1}^{n} N_i^2}\left\{\sum_{i=1}^{n} \frac{y_i^2}{p_i p'_i} - \hat{Y}_{RHC}\right\}$$

Hence the proof. ■

4.4 Midzuno Scheme

This is another unequal probability sampling scheme due to Midzuno (1952). Let $\hat{X}$ be an unbiased estimator of the population total X of the size variable x

under simple random sampling. That is, $\hat{X} = \frac{N}{n}\sum_{i \in s} X_i$. The Midzuno sampling design is defined as

$$P(s) = \begin{cases} \frac{\hat{X}}{X}\frac{1}{\binom{N}{n}} & \text{if } n(s) = n \\ 0 & \text{otherwise} \end{cases} \tag{4.29}$$

The above sampling design can be implemented by using the following sampling method.

To draw a sample of size n, one unit is drawn by probability proportional to size method and from the remaining $(N-1)$ units a simple random sample of size $(n-1)$ will be drawn.

Now we shall prove that the above sampling method will implement the sampling design defined in (4.29).

Let $s = \{i_1, i_2, ..., i_n\}$. The probability of getting the set s as sample is

$$P(s) = \sum_{r=1}^{n} P(A_{i_r}) \tag{4.30}$$

where $P(A_{i_r})$ is the probability of obtaining the set s as sample with r selected in the first draw. It is to be noted that $P(A_{i_r}) = \frac{X_{i_r}}{X}\binom{N-1}{n-1}^{-1}$

Therefore by (4.30), $P(s) = \left[\sum_{i \in s}\frac{X_i}{X}\right]\binom{N-1}{n-1}$

$$= \frac{\hat{X}}{X}\binom{N}{n}^{-1} \qquad \text{(using } X = \sum_{i \in s} X_i\text{)}$$

From this, we infer that the sampling scheme described above implements the sampling design defined in (4.29).

The following theorem gives the first order inclusion probabilities corresponding to the Midzuno sampling design.

Theorem 4.9 Under Midzuno sampling design the first order inclusion probabilities are

$$\pi_i = \frac{N-n}{N-1}\frac{X_i}{X} + \frac{n-1}{N-1}, i = 1, 2, ..., N$$

Proof By definition $\pi_i = \sum_{s \ni i} P(s)$

$$=\binom{N}{n}^{-1}\sum_{s\ni i}\frac{\hat{X}}{X}$$

$$=\binom{N}{n}^{-1}\frac{N}{n}\frac{1}{X}\sum_{s\ni i}\sum_{i\in s}X_i \tag{4.31}$$

Note that (1) the number of subsets of size n containing the label i is $\binom{N-1}{n-1}$ and (2) the number of subsets of size n containing the label j along with label i is $\binom{N-2}{n-2}$. Therefore, by (4.31),

$$\pi_i=\frac{1}{X}\binom{N-1}{n-1}^{-1}\left[\binom{N-1}{n-1}X_i+\binom{N-2}{n-2}(X-X_i)\right]$$

$$=\frac{1}{X}\binom{N-1}{n-1}^{-1}\left[\binom{N-2}{n-2}\left\{\frac{N-n}{n-1}\right\}(X_i+X)\right]$$

$$=\frac{N-n}{N-1}\frac{X_i}{X}+\frac{n-1}{N-1},\ i=1,2,\ldots,N$$

Hence the proof. ■

The following theorem gives the second order inclusion probabilities under Midzuno scheme.

Theorem 4.10 Under Midzuno sampling design, the second order inclusion probabilities are

$$\pi_{ij}=\frac{(N-n)(n-1)}{(N-1)(N-2)}\frac{X_i+X_j}{X}+\frac{(n-1)(n-2)}{(N-1)(N-2)}$$

Proof By definition, $\pi_{ij}=\sum_{s\ni i,j}P(s)$

$$=\frac{N}{n}\frac{1}{X}\sum_{s\ni i,j}\sum_{i\in s}X_i\frac{1}{\binom{N}{n}}$$

Note that the number of subsets of size n containing the labels i and j is $\binom{N-2}{n-2}$ and the number of subsets of size n containing the label k along with labels i and j is $\binom{N-3}{n-3}$.

Therefore $$\pi_{ij}=\frac{1}{X}\frac{1}{\binom{N-1}{n-1}}\left[\binom{N-2}{n-2}(X_i+X_j)+\binom{N-3}{n-3}(X-X_i-X_j)\right]$$

$$=\frac{1}{X}\frac{1}{\binom{N-1}{n-1}}\left[\binom{N-3}{n-3}\left\{\frac{N-n}{n-2}\right\}(X_i+X_j+X)\right]$$

$$=\frac{(N-n)(n-1)}{(N-1)(N-2)}\frac{X_i+X_j}{X}+\frac{(n-1)(n-2)}{(N-1)(N-2)}$$

Hence the proof. ■

Thus we have derived the first and second order inclusion probabilities under Midzuno sampling scheme. These expressions can be used in the Horvitz-Thompson estimator to estimate the population total and derive the variance of the estimator. Midzuno sampling design is one in which the Yates-Grundy estimator of variance is non-negative.

4.5 PPS Systematic Scheme

As in the case of cumulative total method, in probability proportional to size systematic sampling, with each unit a number of numbers equal to its size are associated and the units corresponding to a sample of numbers drawn systematically will be selected as sample. That is, in sampling n units with this procedure, the cumulative totals $T_i, i=1,2,...,N$, are determined and the units corresponding to the numbers $\{r+jk\}, j=0,1,2,...,(n-1)$ are selected, where $k=\frac{T}{n}=\frac{X}{n}$ and r is a random number from 1 to k. This procedure is known as *pps* systematic sampling. The unit U_i is included in the sample, if $T_{i-1}<r+jk\le T_i$ for some value of $j=0,1,2,...,(n-1)$. Since the random number, which determines the sample, is selected from 1 to k and since X_i of the numbers are favourable for inclusion of the ith unit in a sample, the probability π_i of inclusion of the ith population unit is $\frac{nX_i}{X}$ provided $k<X_i$.

It is to be noted that if $\frac{X}{n}$ is not an integer, the sampling interval k can be taken as the integer nearest to $\frac{X}{n}$ and in this case the actual sample size differs from the required sample size. This difficulty can be overcome by selecting the sample in a circular fashion after choosing the random start from 1 to X instead of from 1 to k. Hartley and Rao(1962) have considered *pps* systematic procedure when the units are arranged at random and derived approximate expressions for the variance and estimated variance which are given below.

$$V(\hat{Y}_{HR}) = \frac{1}{n}\sum_{i=1}^{N}\left[\frac{Y_i}{P_i} - Y\right]^2 P_i[1-(n-1)P_i]$$

$$v(\hat{Y}_{HR}) = \frac{1}{n^2(n-1)}\sum_{i=1}^{n}\sum_{i'<i}\left[1-n(p_i+p_{i'})+n\sum_{i''=1}^{n}p_{i''}^2\left[\frac{y_i}{p_i}-\frac{y_{i''}}{p_{i''}}\right]^2\right]$$

It can be shown that even when the units are arranged at random *pps* systematic sampling is more efficient than *ppswr* sampling.

4.6 Problems and Solutions

Problem 4.1 Derive the variance of Desraj ordered estimator when the sample size is two.

Solution When $n = 2$, the estimator $\hat{Y}_{DR}$ can be written as

$$\hat{Y}_{DR} = \frac{1}{2}(t_1 + t_2)$$

$$= \frac{1}{2}\left[y_1\left(\frac{1+p_1}{p_1}\right) + y_2\left(\frac{1-p_1}{p_2}\right)\right]$$

Note that the above estimator can take the values

$$\left[Y_i\left(\frac{1+P_i}{P_i}\right) + Y_j\left(\frac{1-P_i}{P_j}\right)\right], i, j = 1, 2, \ldots, N; i \neq j$$

with respective probabilities $\dfrac{P_iP_j}{1-P_i}$.

Therefore $$E[\hat{Y}_{DR}^2] = \frac{1}{4}\sum_{i=1}^{N}\sum_{\substack{j=1 \\ i\neq j}}^{N}\left[Y_i\left(\frac{1+P_i}{P_i}\right) + Y_j\left(\frac{1-P_i}{P_j}\right)\right]^2 \frac{P_iP_j}{1-P_i}$$

Using the identity, $\sum_{i=1}^{n}\sum_{j=1}^{n}a_{ij} = \sum_{i=1}^{n}a_{ii} + \sum_{i\neq j}^{n}a_{ij}$, we can write

$$E[\hat{Y}_{DR}^2] = \frac{1}{4}\sum_{i=1}^{N}\sum_{j=1}^{N}\left[Y_i\left(\frac{1+P_i}{P_i}\right) + Y_j\left(\frac{1-P_i}{P_j}\right)\right]^2 \frac{P_iP_j}{1-P_i}$$

$$-\frac{1}{4}\sum_{i=1}^{N}\sum_{j=1}^{N}\left[Y_i\left(\frac{1+P_i}{P_i}\right) + Y_i\left(\frac{1-P_i}{P_i}\right)\right]^2 \frac{P_i^2}{1-P_i}$$

On simplification the first and second terms reduce to

$$\frac{1}{4}\sum_{i=1}^{N}\frac{Y_i^2(1+P_i)^2}{P_i(1-P_i)} + \frac{1}{4}[1-\sum_{i=1}^{N}P_i^2]\sum_{j=1}^{N}\frac{Y_j^2}{P_j} + \frac{1}{2}Y^2 - \sum_{i=1}^{N}\frac{Y_i^2}{(1-P_i)}$$

and $\sum_{i=1}^{N} \frac{Y_i^2}{(1-P_i)}$ respectively.

Therefore

$$V(\hat{Y}_{DR}) = \frac{1}{4}\sum_{i=1}^{N} \frac{Y_i^2(1+P_i)^2}{P_i(1-P_i)} + \frac{1}{4}[1-\sum_{i=1}^{N} P_i^2]\sum_{j=1}^{N} \frac{Y_j^2}{P_j} - 2Y^2 - 4\sum_{i=1}^{N} \frac{Y_i^2}{(1-P_i)} + 2Y\sum_{i=1}^{N} Y_i P_i$$

Using $\sum_{i=1}^{N} \frac{Y_j^2}{P_j} = \sum_{i=1}^{N} \left[\frac{Y_i}{P_i} - Y\right]^2 P_i + Y^2$ in the above expression and simplifying the resulting expression we get

$$V(\hat{Y}_{DR}) = \left[1 - \frac{1}{2}\sum_{i=1}^{N} P_i^2\right] \frac{1}{2}\sum_{i=1}^{N} \left[\frac{Y_i}{P_i} - Y\right]^2 P_i - \frac{1}{4}\sum_{i=1}^{N} \left[\frac{Y_i}{P_i} - Y\right]^2 P_i^2$$

Hence the solution. ■

Problem 4.2 A finite population of size N is divided randomly into n groups of equal sizes (assuming the population is a multiple of sample size) and one unit is drawn from each group randomly. Suggest an unbiased estimator for the population total and derive its variance. Compare the resulting variance with the variance of conventional estimator under simple random sampling.

Solution We know that probability proportional to size sampling reduces to equal probability sampling if the units are of the same size. Therefore, the sampling scheme described in the given problem can be viewed as a particular case of random group method. Hence the results given under the random group method can be used for the sampling scheme given above by taking $X_1 = X_2 = ... = X_N = X_0$ (say). In this case, we have

$$P_{ij} = \frac{X_{ij}}{\sum_{k=1}^{N_i} X_{ik}} = \frac{X_0}{N_i X_0}$$

$$= \frac{1}{N_i} = \frac{n}{N}, \; j = 1, 2, ..., N_i; i = 1, 2, ..., n$$

and $P_i = \frac{X_i}{X} = \frac{X_0}{NX_0} = \frac{1}{N}$

Under this set up, by Theorem 4.6 an unbiased estimator of the population total is given by $\hat{Y}_{RHC} = \frac{N}{n}\sum_{i=1}^{n} y_i$.

Further $\sum_{r=1}^{N}\left[\frac{Y_r}{P_r}-Y\right]^2 P_r=\sum_{r=1}^{N}[NY_r-N\bar{Y}]^2\frac{1}{N}$

$$=\sum_{r=1}^{N}N[Y_r-\bar{Y}]^2$$

Substituting this expression in the variance expression given under the Remark (stated below Theorem 4.7) we get

$$V(\hat{Y}_{RHC})=\frac{N^2(N-n)}{Nn}\frac{1}{N-1}\sum_{i=1}^{N}[Y_i-\bar{Y}]^2$$

It may be noted that the variance expression given above is nothing but the variance of the expansion estimator under simple random sampling and in fact the estimator $\hat{Y}_{RHC}=\frac{N}{n}\sum_{i=1}^{n}y_i$ is nothing but the expansion estimator under simple random sampling. ■

Problem 4.3 Show that the Yates-Grundy estimator is non-negative under Midzuno sampling design.

Solution We have seen in Chapter 1, a set of sufficient conditions for the non-negativity of Yates-Grundy estimator are given by

$$\pi_i\pi_j-\pi_{ij}\geq 0, i,j=1,2,\ldots,N; i\neq j$$

Using the expressions given in Theorems 4.9 and 4.10 we have

$$\pi_i\pi_j-\pi_{ij}=\left[\frac{N-n}{N-1}\frac{X_i}{X}+\frac{n-1}{N-1}\right]\left[\frac{N-n}{N-1}\frac{X_j}{X}+\frac{n-1}{N-1}\right]$$

$$-\frac{(N-n)(n-1)}{(N-1)(N-2)}\frac{X_i+X_j}{X}-\frac{(n-1)(n-2)}{(N-1)(N-2)}$$

$$=\left[\left(\frac{N-n}{N-1}\right)^2\frac{X_iX_j}{X^2}+\frac{(N-n)(n-1)}{N-1}\frac{X_i+X_j}{X}\cdot\left[\frac{1}{N-1}-\frac{1}{N-2}\right]\right]$$

$$+\frac{n-1}{N-1}\left[\frac{n-1}{N-1}-\frac{n-2}{N-2}\right]$$

$$=\left(\frac{N-n}{N-1}\right)^2\frac{X_iX_j}{X^2}+\left[\frac{(N-n)(n-1)}{(N-1)^2(N-2)}\left[1-\frac{X_i+X_j}{X}\right]\right]\cdot$$

$$+\frac{(N-n)(n-1)}{(N-1)^2(N-2)}$$

Since the right hand side of the above expression is non-negative, we conclude that the Yates-Grundy estimator is always nonnegative. ■

Problem 4.4 Derive the bias and mean square error of $\frac{N}{n}\sum_{i=1}^{n} y_i$ under probability proportional to size sampling with replacement.

Solution Bias of the estimator is given by

$$B = \frac{N}{n}\sum_{i=1}^{n} E(y_i) - Y$$

$$= \frac{N}{n}\sum_{i=1}^{n}\sum_{j=1}^{N} Y_j P_j - N\overline{Y}$$

$$= \frac{N}{n} n \sum_{j=1}^{N} [Y_j - \overline{Y}] P_j$$

$$= N \sum_{j=1}^{N} [Y_j - \overline{Y}] P_j$$

Consider the difference $\frac{N}{n}\sum_{i=1}^{n} y_i - Y = \frac{N}{n}\sum_{i=1}^{n} (y_i - \overline{Y})$

Squaring both the sides and taking expectation on both the sides we get the mean square error as

$$M = \frac{N^2}{n^2}\left[\sum_{i=1}^{n} E(y_i - \overline{Y})^2 + 2\sum_{i<j}^{N} E(y_i - \overline{Y})(y_j - \overline{Y})\right]$$

$$= \frac{N^2}{n^2}\left[\sum_{i=1}^{n}\sum_{j=1}^{N} (Y_j - \overline{Y})^2 P_j + 2\sum_{i<j}^{N} 0\right]$$

$$= \frac{N^2}{n}\left[\sum_{i=1}^{N} (Y_i - \overline{Y})^2 P_i\right]$$

Cross product terms become zero because units are drawn independently one by one with replacement. ■

Exercises

4.1 Derive the first and second order inclusion probabilities in *PPSWOR* when *n*=2.

4.2 Derive the necessary and sufficient condition for the variance estimator to be non-negative in *PPSWOR* when *n*=2

4.3 Suppose the units in a population are grouped on the basis of equality of their sizes and that each such group has at least n units. Then a sample of *n* units is chosen with *ppswr* from the whole population and repeated units are replaced by units selected with *srswor* from the respective groups.

Suggest an unbiased estimator of the population mean and derive its variance and compare with the usual *PPSWR*.

4.4 If in a sample of three units, drawn with *PPSWR*, only two units are distinct, show that the estimators

$$\frac{1}{3}\left[\frac{y_1}{p_1}+\frac{y_2}{p_2}+\frac{y_1+y_2}{p_1+p_2}\right], \frac{y_1}{1-(1-p_1)^3}+\frac{y_2}{1-(1-p_2)^3}$$

are unbiased for the population total.

Chapter 5

Stratified Sampling

5.1 Introduction

In simple random sampling, it has been seen that the precision of the standard estimator of the population total (mean) depends on two aspects, namely, the sample size and the variability of the character under study (refer the expression (2.6) given in Chapter 2). Therefore in order to get an estimator with increased precision one can increase the sample size. However considerations of cost limit the size of the sample. The other possible way to estimate the population total (mean) with greater precision is to divide the population into several groups each of which is more homogeneous than the entire population and draw sample of predetermined size from each of these groups. The groups into which the population is divided are called strata and drawing sample from each of the strata is called stratified sampling. In stratified sampling, samples are drawn independently from different strata and it is not necessary to use the same sampling design in all strata. Depending on the nature of the strata, different sampling designs can be used in different strata. For example, in the absence of suitable size information, simple random sampling can be used in few strata, whereas probability proportional to size sampling can be used in the remaining strata when size information is available in those strata.

Notations

N : Population size : Number of strata in the population

N_h : Number of units in the stratum $h, h = 1, 2, ..., L$

Y_{hj} : y-value of the jth unit in the stratum $h, j = 1, 2, ..., N_h; h = 1, 2, ..., L$

n_h : Sample size corresponding to the stratum $h, h = 1, 2, ..., L$

Y_h : Stratum total of the stratum $h, h = 1, 2, ..., L$

$\overline{Y}_h$: Stratum mean of the stratum $h, h = 1, 2, ..., L$

y_{hj} : y-value of the jth sampled unit in the stratum

$h, j = 1, 2, ..., N_h; h = 1, 2, ..., L$

$\overline{y}_h$: Stratum sample mean of the stratum $h, h = 1, 2, ..., L$

$$S_h^2 = \frac{1}{N_h - 1} \sum_{j=1}^{N_h} [Y_{hj} - \bar{Y}_h]^2 \, , \; s_h^2 = \frac{1}{n_h - 1} \sum_{j=1}^{n_h} [y_{hj} - \bar{y}_h]^2$$

The following theorems help us to identify unbiased estimator for the population total under different sampling designs and also to obtain their variances.

Theorem 5.1 If $\hat{Y}_h$, $h = 1, 2, ..., L$ is unbiased for the stratum total Y_h of the stratum , then an unbiased estimator for the population total Y is $\hat{Y}_{st} = \sum_{h=1}^{L} \hat{Y}_h$ and its variance is $V(\hat{Y}_{st}) = \sum_{h=1}^{L} V(\hat{Y}_h)$

Proof Since $\hat{Y}_h$ is unbiased for the stratum total Y_h of the stratum h, we have

$$E(\hat{Y}_h) = Y_h \, , h = 1, 2, ..., L$$

Therefore
$$E(\hat{Y}_{st}) = \sum_{h=1}^{L} E(\hat{Y}_h)$$
$$= \sum_{h=1}^{L} Y_h = Y$$

Hence $\hat{Y}_{st} = \sum_{h=1}^{L} \hat{Y}_h$ is unbiased for the population total.

Since samples are drawn independently from different strata, $\mathrm{cov}(\hat{Y}_h, \hat{Y}_k) = 0$ for $h \neq k$.

Therefore
$$V(\hat{Y}_{st}) = \sum_{h=1}^{L} V(\hat{Y}_h) + 2 \sum_{h=1}^{L} \sum_{h<k}^{L} \mathrm{cov}(\hat{Y}_h, \hat{Y}_k)$$
$$= \sum_{h=1}^{L} V(\hat{Y}_h)$$

Hence the proof. ■

Corollary 5.1 If $v(\hat{Y}_h)$ is unbiased for $V(\hat{Y}_h)$, $h = 1, 2, ..., L$, then an unbiased estimator of $V(\hat{Y}_{st})$ is $v(\hat{Y}_{st}) = \sum_{h=1}^{L} v(\hat{Y}_h)$

Proof of this corollary is straight forward and hence omitted.

Corollary 5.2 If simple random sampling is used in all the L strata, then an unbiased estimator of the population total is $\hat{Y}_{st} = \sum_{h=1}^{L} \frac{N_h}{n_h} \sum_{j=1}^{n_h} y_{hj}$.

Proof We know that when a simple random sample of size n is drawn from a population containing N units, $\frac{N}{n}\sum_{i \in s} Y_i$ is unbiased for the population total.

Therefore $\frac{N_h}{n_h}\sum_{j=1}^{n_h} y_{hj}$ is unbiased for the stratum total Y_h , $h = 1, 2, ..., L$. Hence we conclude that $\sum_{h=1}^{L} \frac{N_h}{n_h} \sum_{j=1}^{n_h} y_{hj}$ is unbiased for the population total Y (refer Theorem 5.1). ■

Corollary 5.3 If simple random sampling is used in all the L strata, then

$$V(\hat{Y}_{st}) = \sum_{h=1}^{L} \frac{N_h^2 (N_h - n_h)}{N_h n_h} S_h^2$$

Proof We know that $V(\hat{Y}_h) = \frac{N_h^2 (N_h - n_h)}{N_h n_h} S_h^2$ where $\hat{Y}_h = \frac{N_h}{n_h}\sum_{j=1}^{n_h} y_{hj}$

Therefore by Theorem 5.1 , $V(\hat{Y}_{st}) = \sum_{h=1}^{L} \frac{N_h^2 (N_h - n_h)}{N_h n_h} S_h^2$

Hence the proof. ■

Corollary 5.4 If simple random sampling is used in all the L strata, then an unbiased estimator of $V(\hat{Y}_{st})$ considered in Corollary 5.3 is

$$v(\hat{Y}_{st}) = \sum_{h=1}^{L} \frac{N_h^2 (N_h - n_h)}{N_h n_h} s_h^2$$

Proof We know that under simple random sampling s^2 is unbiased for S^2 (refer Theorem 2.4). Therefore an unbiased estimator of $V(\hat{Y}_h)$ considered in Corollary 5.3 is $v(\hat{Y}_h) = \frac{N_h^2 (N_h - n_h)}{N_h n_h} s_h^2$

Hence by Corollary 5.1, an unbiased estimator of $V(\hat{Y}_{st})$ is

$$v(\hat{Y}_{st}) = \sum_{h=1}^{L} \frac{N_h^2 (N_h - n_h)}{N_h n_h} s_h^2$$

Hence the proof. ■

5.2 Sample Size Allocation

Once the sample size n is fixed next arises the question of deciding the sample size n_h meant for the stratum $h, h = 1, 2, ..., L$. In this section some solutions are given assuming that simple random sampling is used in all the L strata. Two popular allocation techniques are (i) Proportional allocation (ii) Neyman allocation.

Proportional Allocation

Under proportional allocation the number of units to be sampled from the stratum h is made proportional to the stratum size. That is, $n_h \propto N_h$, $h = 1, 2, ..., L$

$$\Rightarrow n_h = kN_h$$

where k is the constant of proportionality. Summing both the sides of the above expression we obtain

$$\sum_{h=1}^{L} n_h = k \sum_{h=1}^{L} N_h$$

$$\Rightarrow n = kN$$

$$\Rightarrow k = \frac{n}{N}$$

Therefore $n_h = \frac{n}{N} N_h$, $h = 1, 2, ..., L$

The following theorem gives an unbiased estimator for the population total and its variance under proportional allocation.

Theorem 5.2 Under proportional allocation , an unbiased estimator for the population total is $\hat{Y}_{st} = \frac{N}{n} \sum_{h=1}^{L} \sum_{j=1}^{n_h} y_{hj}$

and its variance is

$$V(\hat{Y}_{st}) = \frac{N^2 (N-n)}{Nn} \sum_{h=1}^{L} \frac{N_h}{N} S_h^2$$

Proof We know that under proportional allocation

$$n_h = \frac{n}{N} N_h \text{ , } h = 1, 2, ..., L$$

Substituting these values in the expressions given in Corollaries 5.2 and 5.3, we get the required results after simplification. ■

The above discussion gives the sample sizes under proportional allocation when the total sample size is known in advance and it does not take into account the cost involved under the allocation. Normally cost will always be a constraint in the organisation of any sample survey. Therefore it is of interest to consider proportional allocation for a given cost. Let c_h , $h = 1, 2, ..., L$ be the cost of collecting information from a unit in stratum *h*. (These costs can differ substantially between strata. For example, information from large establishments can be obtained cheaply if we mail them questionnaire, whereas small establishments may have to be personally contacted in order to get reliable data). Therefore the total cost of the survey can be taken as

$$C = C_0 + \sum_{h=1}^{L} c_h n_h \tag{5.1}$$

where C_0 is the fixed cost. When the sample size n_h is proportional to the stratum size, we have

$$n_h = kN_h \, , h = 1, 2, \ldots, L \tag{5.2}$$

where k is the constant of proportionality. Summing both the sides of (5.2) with respect to h after multiplying by c_h, we get

$$\sum_{h=1}^{L} c_h n_h = \sum_{h=1}^{L} c_h N_h$$

Using (5.1) in the above expression we get

$$C - C_0 = k \sum_{h=1}^{L} c_h N_h \quad \Rightarrow k = \frac{C - C_0}{\sum_{h=1}^{L} c_h N_h}$$

Therefore the proportional allocation for a given cost is given by

$$n_h = \frac{C - C_0}{\sum_{h=1}^{L} c_h N_h} N_h \tag{5.3}$$

Summing both sides with respect to h , $h = 1, 2, \ldots, L$ we get the total sample size as

$$n = \frac{C - C_0}{\sum_{h=1}^{L} c_h N_h} N \tag{5.4}$$

Under the above allocation the variance of the estimator defined in Theorem 5.1 is

$$\frac{1}{C - C_0} \sum_{h=1}^{L} N_h S_h^2 \sum_{h=1}^{L} c_h n_h - \sum_{h=1}^{L} N_h S_h^2 \tag{5.5}$$

Optimum Allocation

The proportional allocations described above do not take into account any factor other than strata sizes. They completely ignore the internal structure of strata like within stratum variability etc., and hence it is desirable to consider an allocation scheme which takes into account these aspects. In this section two allocation schemes which minimise the variance of the estimator are considered. Since minimum variance is an optimal property, these allocations are called

"Optimum allocations". Note that under simple random sampling the variance of $\hat{Y}_{st} = \sum_{h=1}^{L} \hat{Y}_h$ can be expressed as

$$V(\hat{Y}_{st}) = \sum_{h=1}^{L} \frac{N_h^2 (N_h - n_h)}{N_h n_h} S_h^2$$

$$= \sum_{h=1}^{L} \frac{N_h^2 S_h^2}{n_h} - \sum_{h=1}^{L} N_h S_h^2$$

Global minimisation of the above variance with respect to $n_1, n_2, \ldots, n_L$ does not yield a non-trivial solution (see what happens when the first order partial derivatives with respect to $n_1, n_2, \ldots, n_L$ are equated to zero). Therefore in order to get non-trivial solutions for $n_1, n_2, \ldots, n_L$, we resort to conditional minimisation. Two standard conditional minimisation techniques are (i) Minimising the variance for a given cost and (ii) Minimising the variance for a given sample size. The solution given by the latter will be referred to as "Neyman optimum allocation" and the former allocation will be referred to as "Cost optimum allocation". The expressions for the sample sizes under the two types of allocations mentioned above are derived below.

(i) Neyman optimum allocation

As mentioned above under Neyman allocation, the variance of the estimator will be minimised by fixing the total sample size. That is, we need the values of $n_1, n_2, \ldots, n_L$ which minimise

$$\sum_{h=1}^{L} \frac{N_h^2 S_h^2}{n_h} - \sum_{h=1}^{L} N_h S_h^2$$

subject to the condition $\sum_{h=1}^{L} n_h = n$.

To solve the above problem consider the function

$$\phi = \sum_{h=1}^{L} \frac{N_h^2 S_h^2}{n_h} - \sum_{h=1}^{L} N_h S_h^2 + \lambda \left\{ \sum_{h=1}^{L} n_h - n \right\} \tag{5.6}$$

where λ is the Lagrangian multiplier. Differentiating the above function partially with respect to n_h and equating the derivatives to zero we get

$$-\frac{N_h^2 S_h^2}{n_h^2} + \lambda = 0, \quad h = 1, 2, \ldots, L$$

$$n_h = \frac{N_h S_h}{\sqrt{\lambda}} \tag{5.7}$$

Differentiating the function ϕ with respect to λ and equating the derivative to zero we get

$$\sum_{h=1}^{L} n_h = n \tag{5.8}$$

Summing both the sides of (5.7) with respect to h from 1 to L, we get

$$\sum_{h=1}^{L} n_h = \frac{\sum_{h=1}^{L} N_h S_h}{\sqrt{\lambda}} \Rightarrow n = \frac{\sum_{h=1}^{L} N_h S_h}{\sqrt{\lambda}}$$

Therefore $\sqrt{\lambda} = \frac{\sum_{h=1}^{L} N_h S_h}{n}$ (5.9)

Using (5.9) in (5.7) we get $n_h = \frac{N_h S_h}{\sum_{h=1}^{L} N_h S_h} n$ (5.10)

The expression given in (5.10) can be used to calculate the sample sizes for different strata. It can be seen that the matrix of second order partial derivatives is positive definite for the values satisfying (5.7) and (5.8). Therefore we conclude that the values yielded by (5.10) minimise the variance of the estimator for the given sample size.

Under the above allocation the variance of the estimator reduces to

$$\frac{1}{n}\left\{\sum_{h=1}^{L} \frac{N_h S_h}{n_h}\right\} - \sum_{h=1}^{L} N_h S_h^2 \tag{5.11}$$

This expression is obtained by using (5.10) in $V(\hat{Y}_{st})$ under simple random sampling.

(ii) Cost Optimum Allocation

Under cost-optimum allocation the sample sizes are determined by minimising the variance of the estimator by fixing the total cost of the survey. As in the case of proportional allocation for a given cost, the total cost of the survey can be taken as $C = C_0 + \sum_{h=1}^{L} c_h n_h$ where C_0 is the fixed cost and c_h is the cost per unit in stratum h, $h = 1, 2, \ldots, L$.

Define $\phi = \sum_{h=1}^{L} \frac{N_h^2 S_h^2}{n_h} - \sum_{h=1}^{L} N_h S_h^2 + \lambda\left\{C_0 + \sum_{h=1}^{L} c_h n_h - C\right\}$ (5.12)

Differentiating the above function partially with respect to n_h and equating the derivatives to zero we get

$$-\frac{N_h^2 S_h^2}{n_h^2} + \lambda\ c_h = 0\ , h = 1, 2, \ldots, L$$

$$n_h = \frac{N_h S_h}{\sqrt{\lambda}\sqrt{c_h}} \tag{5.13}$$

Differentiating the function ϕ with respect to λ and equating the derivative to zero we get

$$C_0 + \sum_{h=1}^{L} c_h n_h = C$$

$$\sum_{h=1}^{L} c_h n_h = C - C_0 \tag{5.14}$$

Summing both the sides of (5.7) with respect to h from 1 to L, we get

$$\sum_{h=1}^{L} c_h n_h = \frac{\sum_{h=1}^{L} N_h S_h \sqrt{c_h}}{\sqrt{\lambda}}$$

$$\sqrt{\lambda} = \frac{\sum_{h=1}^{L} N_h S_h \sqrt{c_h}}{\sum_{h=1}^{L} c_h n_h}$$

$$= \frac{\sum_{h=1}^{L} N_h S_h \sqrt{c_h}}{(C - C_0)} \quad \text{(using (5.14))}$$

Using this expression in (5.13) we get

$$n_h = \frac{\frac{N_h S_h}{\sqrt{c_h}}(C - C_0)}{\sum_{h=1}^{L} N_h S_h \sqrt{c_h}}\ , h = 1, 2, \ldots, L \tag{5.15}$$

The expression given above gives the optimum allocation of the sample for a given cost. It can be shown that the matrix of the second order partial derivatives is positive definite. Hence we conclude that this allocation minimises the variance for a given cost.

The expression given in (5.15) leads to the following conclusions. In a given stratum, take a large sample if : (1) the stratum size is larger; (2) the stratum has more internal variation with respect to the variable under study; (3) sampling is cheaper in the stratum.

Summing both sides of the equation (5.15) with respect to h from 1 to L, we get the total sample size under the cost-optimum allocation as

$$n=\frac{(C-C_0)\sum_{h=1}^{L}\frac{N_hS_h}{\sqrt{c_h}}}{\sum_{h=1}^{L}N_hS_h\sqrt{c_h}} \tag{5.16}$$

The variance of the conventional estimator under the cost-optimum allocation reduces to

$$\frac{\left\{\sum_{h=1}^{L}N_hS_h\sqrt{c_h}\right\}^2}{C-C_0}-\sum_{h=1}^{L}N_hS_h^2 \tag{5.17}$$

This expression is obtained by using (5.15) in $V(\hat{Y}_{st})$ under simple random sampling.

The following theorem compares the variance of conventional estimator under simple random sampling, proportional allocation for a given sample size and optimum allocation for a given sample size.

Theorem 5.3 Let V_{ran}, V_{prop} and V_{opt} be the variances of the usual estimators under simple random sampling, proportional allocation and optimum allocation for a given sample size. If N_h is large then

$$V_{ran}\le V_{prop}\le V_{opt}$$

Proof We know that under simple random sampling, the variance of the conventional estimator for the population total is

$$V_{ran}=\frac{N^2(N-n)}{Nn}S^2 \tag{5.18}$$

Note that $(N-1)S^2=\sum_{h=1}^{L}\sum_{j=1}^{N_h}[Y_{hj}-\bar{Y}]^2$

$$=\sum_{h=1}^{L}\sum_{j=1}^{N_h}[Y_{hj}-\bar{Y}_h]^2+\sum_{h=1}^{L}N_h[\bar{Y}_h-\bar{Y}]^2$$

$$=\sum_{h=1}^{L}\sum_{j=1}^{N_h}(N_h-1)S_h^2+\sum_{h=1}^{L}N_h[\bar{Y}_h-\bar{Y}]^2 \tag{5.19}$$

Therefore $S^2\cong\sum_{h=1}^{L}W_hS_h^2+\sum_{h=1}^{L}W_h[\bar{Y}_h-\bar{Y}]^2$ (5.20) where $W_h = \dfrac{N_h}{N}$.

This is obtained by using the fact that N_h and hence N is large.

Using (5.20) in (5.18) we get

$$V_{ran} = \frac{N^2(N-n)}{Nn}\left[\sum_{h=1}^{L} W_h S_h^2 + \sum_{h=1}^{L} W_h [\overline{Y}_h - \overline{Y}]^2\right]$$

$$= V_{prop} + \frac{N^2(N-n)}{Nn}\left[\sum_{h=1}^{L} W_h [\overline{Y}_h - \overline{Y}]^2\right] \quad \text{(By Theorem 5.2)}$$

Therefore $V_{ran} \geq V_{prop}$ (*)

By expression (5.11) we have

$$V_{opt} = \frac{1}{n}\left\{\sum_{h=1}^{L} N_h S_h\right\}^2 - \sum_{h=1}^{L} N_h S_h^2$$

Therefore

$$V_{prop} - V_{opt} = \frac{N^2(N-n)}{Nn}\sum_{h=1}^{L} W_h S_h^2 - \frac{1}{n}\left\{\sum_{h=1}^{L} N_h S_h\right\}^2 + \sum_{h=1}^{L} N_h S_h^2$$

$$= \frac{(N-n)}{n}\sum_{h=1}^{L} N_h S_h^2 - \frac{1}{n}\left\{\sum_{h=1}^{L} N_h S_h\right\}^2 + \sum_{h=1}^{L} N_h S_h^2$$

$$= \frac{(N-n)}{n}\sum_{h=1}^{L} N_h S_h^2 - \frac{N^2}{n}\overline{S}^2 \quad \text{where } \overline{S} = \frac{1}{N}\sum_{h=1}^{L} N_h S_h$$

$$= \frac{N}{n}\sum_{h=1}^{L} N_h (S_h - \overline{S})^2 \tag{5.21}$$

Therefore $V_{prop} \geq V_{opt}$ (**)

The result follows from (*) and (**).■

Note From (5.21) we have $V_{prop} - V_{opt} = \dfrac{N}{n}\sum_{h=1}^{L} N_h (S_h - \overline{S})^2$

Therefore

$$V_{ran} = V_{opt} + \frac{N}{n}\sum_{h=1}^{L} N_h (S_h - \overline{S})^2 + \left[\sum_{h=1}^{L} W_h [\overline{Y}_h - \overline{Y}]^2\right]$$

This expression leads to the conclusion that as we change from simple random sampling to optimum allocation with fixed sample size, considerable amount of precision can be gained by forming the strata such that variance between means and variances are large.

5.3 Comparison with Other Schemes

(1) Comparison under populations with linear trend

Suppose that the population is divided (assuming that $N = nk$, n and k being integers) into n strata where the stratum h contains units with labels

$$G_h = \{(h-1)k + j,\ j = 1, 2, \ldots, k\},\ h = 1, 2, \ldots, n$$

and one unit is selected from each stratum randomly to get a sample of size *n*.

Under the above stratification-sampling scheme an unbiased estimator for the population total is given by $\hat{Y}_0 = \frac{k}{1}\sum_{h=1}^{n} y_{h1} = k\sum_{h=1}^{n} y_{h1}$

This estimator is derived from Corollary 5.2, by taking $L = n, N_h = k, n_h = 1$. On applying Corollary 5.3, we obtain the variance of the above estimator as

$$\sum_{h=1}^{n} \frac{k^2(k-1)}{k} \frac{1}{k-1} \sum_{j=1}^{k} [Y_{hj} - \overline{Y}_h]^2$$

which reduces to $k\sum_{h=1}^{n}\sum_{j=1}^{k}[Y_{hj} - \overline{Y}_h]^2$ (5.22)

When the population values are modeled by the relation,

$$Y_i = a + bi, i = 1, 2, \ldots, N$$

under the stratification scheme described earlier, we have

$$Y_{hji} = a + b[(h-1)k + j]$$

Therefore $\overline{Y}_h = \frac{1}{k}\sum_{j=1}^{k}\{a + b[(h-1)k + j]\}$

$$= a + b\left\{(h-1)k + \frac{k(k+1)}{2k}\right\}$$

$$\Rightarrow Y_{hj} - \overline{Y}_h = b\left\{j - \frac{(k+1)}{2}\right\}$$

Squaring both the sides and summing with respect to *j* from 1 to *k*, we get

$$\sum_{j=1}^{k}[Y_{hj} - \overline{Y}_h]^2 = b\sum_{j=1}^{k}\left\{j^2 + \frac{(k+1)^2}{4} - (k+1)j\right\}$$

$$= b^2\left\{\frac{k(k+1)(2k+1)}{6} + \frac{k(k+1)^2}{4} - \frac{k^2(k+1)^2}{2}\right\}$$

$$= b^2\,\frac{k(k^2-1)}{12}$$

Using this in (5.22) we get the variance of the estimator $\hat{Y}_0$ as

$$V(\hat{Y}_0) = b^2 \frac{nk^2(k^2-1)}{12} \tag{5.23}$$

Already we have seen in Chapter 3, for populations exhibiting linear trend,

$$V(\hat{Y}_{srs}) = b^2 \frac{n^2k^2(k-1)(nk+1)}{12} \tag{5.24}$$

$$V(\hat{Y}_{LSS}) = b^2 \frac{n^2k^2(k^2-1)}{12} \tag{5.25}$$

Denoting the variances given in (5.23),(5.24) and (5.25) by V_{st}, V_{ran}, V_{sys}, we obtain $V_{st} \le V_{ran} \le V_{sys}$.

From the above inequality, we conclude that the stratification -estimation scheme described in this section is better than both simple random sampling and systematic sampling for populations exhibiting linear trend.

(2) Comparison under Autocorrelated trend

Assuming that the N population values are the realized values of N random variables having a joint distribution such that

$$E_M[Y_i] = \mu,\ E_M[Y_i - \mu]^2 = \sigma^2 \text{ and}$$

$$E_M[Y_i - \mu][Y_{i+u} - \mu] = \rho_u \sigma^2 \text{ where } \rho_u \ge \rho_v \text{ whenever } u < v,$$

we have proved that (Theorem 3.8)

$$E_M V(\hat{Y}_{srs}) = \frac{\sigma^2(k-1)N^2}{nk}\left[1 - \frac{2}{N(N-1)}\sum_{u=1}^{N-1}(N-u)\rho_u\right] \tag{5.26}$$

The expected variance of the estimator $\hat{Y}_0$ under the above model is given by

$$E_M[V(\hat{Y}_0)] = E_M\left\{k\sum_{h=1}^{n}\sum_{j=1}^{k}[Y_{hj} - \overline{Y}_h]^2\right\} \tag{5.27}$$

$$= \frac{\sigma^2(k-1)N^2}{nk}\left[1 - \frac{2}{k(k-1)}\sum_{u=1}^{k-1}(k-u)\rho_u\right] \tag{5.28}$$

(refer Theorem 3.21)

Define $$L(j) = \frac{2}{j(j-1)}\sum_{u=1}^{j-1}(j-u)\rho_u,\ j = 2, 3, \ldots \tag{5.29}$$

Then $$E_M V(\hat{Y}_{srs}) = \frac{\sigma^2(k-1)N^2}{nk}[1 - L(nk)] \tag{5.30}$$

and $$E_M V(\hat{Y}_0) = \frac{\sigma^2(k-1)N^2}{nk}[1 - L(k)] \tag{5.31}$$

Thus in order to prove $E_M V(\hat{Y}_0) \le E_M V(\hat{Y}_{srs})$ it is enough to show that

$$L(nk) \le L(k) \tag{5.32}$$

Consider the difference

$$L(j)-L(j+1)=\frac{2}{j(j^2-1)}\sum_{u=1}^{j}(j+1-2u)\rho_u \tag{5.33}$$

If S stands for the summation term in the right hand side of (5.33), grouping together the terms equidistant from the beginning and end, S can be written as

$$S=\sum_{u=1}^{m}[2m+1-2u][\rho_u-\rho_{2m+1-u}] \quad \text{if } j=2m \text{ is even}$$

$$S=\sum_{u=1}^{m}[2m+2-2u][\rho_u-\rho_{2m+2-u}] \quad \text{if } j=2m+1 \text{ is odd}$$

Since $\rho_i \geq \rho_{i+1}$ for all i, every term in S is non-negative. Therefore S is non-negative. Hence we conclude that L is a non-decreasing function. Therefore $L(nk)\leq L(k)$. This leads to the conclusion that the average variance of the conventional estimator under simple random sampling is larger than the average variance of the estimator $\hat{Y}_0$ introduced in this section. However no such general result can be proved about the efficiency of systematic sampling relative to simple random sampling or stratified sampling unless further restrictions are imposed on the correlations ρ_u. The following theorem is due to Cochran (1946).

Theorem 5.4 If $\rho_i \geq \rho_{i+1} \geq 0, i=1,2,...,N-2$, $\partial_i^2=\rho_{i+2}+\rho_i-2\rho_{i+1}\geq 0$, and $i=1,2,...,N-2$ then $E_M[V(\hat{Y}_{LSS})]\leq E_M[V(\hat{Y}_0)]\leq E_M[V(\hat{Y}_{srs})]$.

Furthermore, unless $\partial_i^2=0, i=1,2,...,N-3.E_M[V(\hat{Y}_{LSS})]\leq E_M[V(\hat{Y}_{LSS})]$

Proof As $\partial_i^2\geq 0$, we have $\rho_{i+2}+\rho_i-2\rho_{i+1}\geq 0, i=1,2,...,N-2$

By induction it can be shown that $\rho_{i+c+1}-\rho_{i+c}\geq \rho_{i+c}-\rho_i$ for any integer c.

Hence for any integer $a,c>0$ we have

$$\sum_{i=a}^{i+c+1}\rho_{i+c+1}-\rho_{i+c}\geq \sum_{i=a}^{a+c-1}\rho_{i+c}-\rho_i$$

which gives $\rho_{a+2c}+\rho_a-2\rho_{a+c}\geq 0$ (5.34)

Consider the difference

$$E_M[V(\hat{Y}_0)]-E_M[V(\hat{Y}_{LSS})]=$$

$$\frac{2\sigma^2(k-1)N^2}{Nnk^2(k-1)}\left[\sum_{u=1}^{nk-1}(nk-u)\rho_u-k^2\sum_{u=1}^{n-1}(n-u)\rho_{ku}-n\sum_{u=1}^{k-1}(k-u)\rho_u\right] \tag{5.35}$$

Note that $\sum_{u=1}^{nk-1}(nk-u)\rho_u=\sum_{i=1}^{k}\sum_{j=0}^{n-1}[nk-(i+jk)]\rho_{i+jk}$

$$=\sum_{i=1}^{k-1}\sum_{j=0}^{n-1}[nk-(i+jk)]\rho_{i+jk}+\sum_{j=1}^{n-1}(n-j)\rho_{jk}$$

$$=\sum_{i=1}^{k-1}\sum_{j=1}^{n-1}(n-j)(k-i)\rho_{i+jk}+k\sum_{j=1}^{n-1}(n-j)\rho_{jk}+$$

$$\sum_{i=1}^{k-1}\sum_{j=0}^{n-2}i(n-j-i)\rho_{jk+i}+n\sum_{i=1}^{k-1}(k-i)\rho_i$$

(5.36)

Since $\sum_{i=1}^{k-1}\sum_{j=0}^{n-2}i(n-j-i)\rho_{jk+i}=\sum_{i=1}^{k-1}\sum_{j=1}^{n-i}i(n-j)\rho_{jk-(k-i)}$

$$=\sum_{i=1}^{k-1}\sum_{j=1}^{n-i}(k-i)(n-j)\rho_{jk-i} \tag{5.37}$$

The expression inside square braces of (5.35) can be written as

$$\sum_{i=1}^{k-1}\sum_{j=1}^{n-i}(k-i)(n-j)\rho_{jk+i}+\sum_{i=1}^{k-1}\sum_{j=1}^{n-i}(k-i)(n-j)\rho_{jk-i}-k(k-1)\sum_{j=1}^{n-1}\rho_{jk}$$

which is equal to $\sum_{i=1}^{k-1}\sum_{j=1}^{n-i}(k-i)(n-j)[\rho_{jk+i}+\rho_{jk-i}-2\rho_{jk}]$.

By (5.34) this is clearly non-negative. Therefore $E_M[V(\hat{Y}_0)]\le E_M[V(\hat{Y}_{LSS})]$. Further from (5.38), it can be seen that the above inequality will be strict if and only if $\partial_i=0, i=1,2,...,N-1$. Hence the proof. ■

5.4 Problems and Solutions

Problem 5.1 A sampler has two strata with relative sizes $W_1=\frac{N_1}{N}$ and $W_2=\frac{N_2}{N}$. He believes that S_1, S_2 can be taken as equal. For a given cost $C=c_1n_1+c_2n_2$, show that (assuming N_h is large)

$$\left[\frac{V_{prop}}{V_{opt}}\right]=\frac{[W_1c_1+W_2c_2]}{[W_1\sqrt{c_1}+W_2\sqrt{c_2}]^2}$$

Solution When N_h is large, $V(\hat{Y}_{st})=\sum_{h=1}^{L}\left\{N_h^2\frac{N_h-n_h}{N_hn_h}\right\}S_h^2$.

$$=\sum_{h=1}^{L}\left\{N_h^2\left[\frac{1}{n_h}-\frac{1}{N_h}\right]\right\}S_h^2$$

$$=\sum_{h=1}^{L}\frac{N_h^2}{n_h}S_h^2 \tag{5.38}$$

For the given cost, under proportional allocation we have

$$n_h=\frac{CN_h}{c_1N_1+c_2N_2},\ h=1,2$$

This expression is obtained from (5.3) by taking $C_0=0$ and $L=2$. Substituting these values in (5.38) we get

$$V_{prop}=\frac{N_1^2S_1^2}{\dfrac{CN_1}{c_1N_1+c_2N_2}}+\frac{N_2^2S_2^2}{\dfrac{CN_2}{c_1N_1+c_2N_2}}$$

$$=\frac{c_1N_1+c_2N_2}{C}\left\{\frac{N_1^2S_1^2}{N_1}+\frac{N_2^2S_2^2}{N_2}\right\}$$

$$=\frac{c_1N_1+c_2N_2}{C}S^2N \tag{5.39}$$

The above expression is obtained by taking $N_1+N_2=N$ and $S_1=S_2$. For the given cost, under optimum allocation we have

$$n_h=\frac{\dfrac{CN_hS_h}{\sqrt{c_h}}}{N_1S_1\sqrt{c_1}+N_2S_2\sqrt{c_2}},\ h=1,2$$

This expression is obtained from (5.15) by taking $C_0=0$ and $L=2$. The variance of standard estimator can be obtained from (5.38) by substituting the sample size values given above. It turns out to be

$$V_{opt}=N_1^2S_1^2\frac{\dfrac{CN_1S_1}{\sqrt{c_1}}}{N_1S_1\sqrt{c_1}+N_2S_2\sqrt{c_2}}+N_2^2S_2^2\frac{\dfrac{CN_2S_2}{\sqrt{c_2}}}{N_1S_1\sqrt{c_1}+N_2S_2\sqrt{c_2}}$$

$$=\frac{[N_1S_1\sqrt{c_1}+N_2S_2\sqrt{c_2}]}{C}[N_1S_1\sqrt{c_1}+N_2S_2\sqrt{c_2}]$$

$$=\frac{[N_1S_1\sqrt{c_1}+N_2S_2\sqrt{c_2}]^2}{C}S^2 \tag{5.40}$$

Therefore by (5.39) and (5.40) we get

$$\frac{V_{prop}}{V_{opt}}=\frac{[W_1c_1+W_2c_2]}{[W_1\sqrt{c_1}+W_2\sqrt{c_2}]^2}$$

Hence the solution. ■

Problem 5.2 With two strata, a sampler would like to have $n_1 = n_2$ for administrative convenience, instead of using the values given by the Neyman allocation. If V and V_{opt} denote the variances given by the $n_1 = n_2$ and the Neyman allocations, respectively, show that the fractional increase in variance $\frac{V - V_{opt}}{V_{opt}} = \left[\frac{r-1}{r+1}\right]^2$ where $r = \frac{n_1}{n_2}$ as given by the Neyman allocation. Assume that N_1 and N_2 is large.

Solution Under equal allocation we have $n_1 = n_2 = \frac{n}{2}$. Substituting this in (5.38) we get (with $n = 2$)

$$V = \left[\frac{2}{n}\right]\left[N_1^2 S_1^2 + N_2^2 S_2^2\right] \tag{5.41}$$

Under Neyman allocation we have

$$n_1 = \frac{N_1 S_1}{N_1 S_1 + N_2 S_2} n \text{ and } n_2 = \frac{N_2 S_2}{N_1 S_1 + N_2 S_2} n$$

Substituting these values in (5.38) we get

$$V_{opt} = \frac{1}{n}\left[N_1 S_1 + N_2 S_2\right]^2 \tag{5.42}$$

By the definition of r, we have $r = \frac{N_1 S_1}{N_2 S_2}$. Using this in V and V_{opt} given in (5.41) and (5.42) we get

$$V = \frac{2}{n} N_2^2 S_2^2 (r^2 + 1) \tag{5.43}$$

$$V_{opt} = \frac{N_2^2 S_2^2}{n}(r+1)^2 \tag{5.44}$$

$$\text{Therefore } \frac{V - V_{opt}}{V_{opt}} = \frac{\left[\frac{2}{n} N_2^2 S_2^2 (r^2 - 1) - \frac{1}{n}\frac{N_2^2 S_2^2}{n}(r+1)^2\right]}{\frac{1}{n}\frac{N_2^2 S_2^2}{n}(r+1)^2}$$

$$= \frac{(r-1)^2}{(r+1)^2}$$

Hence the solution. ■

Problem 5.3 If there are two strata and if ϕ is the ratio of the actual $\frac{n_1}{n_2}$ to the Neyman optimum $\frac{n_1}{n_2}$, show that whatever be the values of N_1, N_2, S_1 and

S_2, the ratio of $\frac{V_{opt}}{V}$ is never less than $\frac{4\phi}{(1+\phi)^2}$ when N_1 and N_2 are large. Here V_{opt} is the variance of usual estimator under Neyman optimum allocation and V is the variance under actual allocation.

Solution By (5.8), $V = \frac{N_1^2}{n_1}S_1^2 + \frac{N_2^2}{n_2}S_2^2$ and by (5.42),

$$V_{opt} = \frac{1}{n}[N_1S_1 + N_2S_2]^2$$

Therefore $$\frac{V_{opt}}{V} = \frac{\frac{1}{n}[N_1S_1 + N_2S_2]^2}{\frac{N_1^2S_1^2}{n_1} + \frac{N_2^2S_2^2}{n_2}} \tag{5.45}$$

Under Neyman allocation,

$$\frac{n_1}{n_2} = \frac{N_1S_1}{N_2S_2} \quad \text{(refer Problem 5.2)}$$

$$\phi = \frac{N_2S_2n_1}{N_1S_1n_2} \tag{5.46}$$

$$\frac{V_{opt}}{V} = \frac{\frac{1}{n}\left\{1 + \frac{N_2S_2}{N_1S_1}\right\}^2}{\frac{1}{n_1} + \frac{N_2^2S_2^2}{N_1^2S_1^2n_2}}$$

The above expression is obtained by dividing both the numerator and denominator of (5.45) by $N_1^2S_1^2$. Substituting the value of ϕ given in (5.45) in the above expression, we get

$$\frac{V_{opt}}{V} = \frac{\frac{1}{n}\left\{1 + \phi\frac{n_2}{n_1}\right\}^2}{\frac{1}{n_1} + \phi^2\frac{n_2^2}{n_1^2n_2}}$$

$$= \frac{\frac{1}{n}(n_1 + n_2\phi)^2}{n_1 + n_2\phi^2} \tag{5.47}$$

Replacing n by $n_1 + n_2$ and $(n_1 + n_2\phi)^2$ by $(n_1 - n_2\phi)^2 + 4n_1n_2\phi$, the above ratio can be expressed as $\frac{V_{opt}}{V} = \frac{(n_1 - n_2\phi)^2 + 4n_1n_2\phi}{(n_1 - n_2\phi)^2 + n_1n_2(1+\phi)^2}$. We know that

$\frac{x+a}{x+b} \geq \frac{a}{b}$ whenever $x \geq 0$. Since $4n_1n_2\phi \geq n_1n_2(1+\phi)^2$ is always true, we conclude that $\frac{V_{opt}}{V} \geq \frac{4\phi}{(1+\phi)^2}$. Hence the solution. ■

Problem 5.4 If the cost function is of the form $C = C_0 + \sum_{h=1}^{L} t_h\sqrt{n_h}$, where C_0 and t_h are known numbers, show that in order to minimize the variance of the estimator for fixed total cost n_h must be proportional to $\left\{\frac{N_h^2 S_h^2}{t_h}\right\}^{\frac{2}{3}}$

Solution To find the desired values of n_h, we must minimize the function

$$\phi = \sum_{h=1}^{L}\left\{N_h^2\left[\frac{1}{n_h} - \frac{1}{N_h}\right]\right\}S_h^2 + \lambda\left\{C_0 + \sum_{h=1}^{L} t_h\sqrt{n_h} - C\right\}$$

where λ is the Lagrangian multiplier.

Differentiating partially the above function with respect to n_h and equating the derivatives to zero we get

$$-\frac{N_h^2 S_h^2}{n_h^2} + \frac{t_h}{2\sqrt{n_h}} = 0, h = 1, 2, ..., L$$

$$n_h^{3/2} = \frac{N_h^2 S_h^2}{\lambda t_h}$$

$$n_h = \frac{\left[\frac{N_h^2 S_h^2}{t_h}\right]^{2/3}}{\lambda^{2/3}} \tag{5.48}$$

Differentiating partially with respect to λ and setting the derivative equal to zero we get

$$C = C_0 + \sum_{h=1}^{L} t_h\sqrt{n_h} \tag{5.49}$$

From (5.48) we have, $\sum_{h=1}^{L} t_h\sqrt{n_h} = \frac{\sum_{h=1}^{L}\left\{\frac{N_h^2 S_h^2}{t_h}\right\}^{1/3}}{\lambda^{1/3}}$

$$C - C_0 = \frac{\sum_{h=1}^{L}\left\{\frac{N_h^2 S_h^2}{t_h}\right\}^{1/3}}{\lambda^{1/3}} \quad \text{(using (5.49))}$$

$$\lambda^{1/3} = \frac{\sum_{h=1}^{L}\left\{\frac{N_h^2 S_h^2}{t_h}\right\}^{1/3}}{C - C_0}$$

Substituting this value in (5.48) we get

$$n_h = \frac{(C - C_0)^2 \left[\frac{N_h^2 S_h^2}{t_h}\right]^{2/3}}{\sum_{h=1}^{L}\left[\frac{N_h^2 S_h^2}{t_h}\right]^{1/3} t_h}$$

It can be seen that for these values of n_h, the matrix of second order partial derivatives becomes positive definite. Therefore we conclude that the above values of n_h minimize the variance of the estimator for a given cost. ■

Problem 5.5 In a population consisting of a linear trend, show that a systematic sample is less precise than stratified random sample with strata of size $2k$ and two units per stratum if $n > \frac{4k+2}{k+1}$ when the first stratum contains first set of $2k$ units, second stratum contains second set of $2k$ units in the population and so on.

Solution We know that under systematic sampling in the presence of linear trend, the variance of the usual estimator is

$$V[\hat{Y}_{LSS}] = \frac{N^2 \beta^2 (k^2 - 1)}{12} \tag{5.50}$$

Under the stratification scheme described above, we have $L = \frac{n}{2}$, $N_h = 2k$ for $h = 1, 2, \ldots, \frac{n}{2}$ and the labels of units included in the stratum h are given by

$$G_h = \{2(h-1)k + j,\ j = 1, 2, \ldots, 2k\},\ h = 1, 2, \ldots, \frac{n}{2}$$

Therefore we have $Y_{hj} = \alpha + \beta\{2(h-1)k + j\},\ j = 1, 2, \ldots, 2k,\ h = 1, 2, \ldots, \frac{n}{2}$

Hence $\overline{Y}_h = \frac{1}{2k}\sum_{j=1}^{2k} \alpha + \beta\{2(h-1)k + j\}$

$$= \alpha + \beta\left[2(h-1)k + \frac{2k+1}{2}\right]$$

$$Y_{hj} - \overline{Y}_h = \beta\left[j - \frac{2k+1}{2}\right]$$

Squaring and summing we get,

$$S_h^2 = \frac{1}{2k-1}\sum_{j=1}^{2k}\beta^2\left[j - \frac{2k+1}{2}\right]^2$$

$$= \beta^2 \frac{k(2k+1)}{6}$$

Since $N_h = 2k, n_h = 2$ and $S_h^2 = \beta^2 \frac{k(2k+1)}{6}$, we obtain the variance of the estimator as

$$V_{st} = \sum_{h=1}^{L}\left\{N_h^2\left[\frac{1}{n_h} - \frac{1}{N_h}\right]\right\}S_h^2$$

$$= \beta^2 \frac{k^2(k-1)n(2k+1)}{6} \tag{5.51}$$

Comparing (5.50) and (5.51) we infer that systematic sampling is less precise than stratified sampling if $n > \frac{4k+2}{k+1}$. Hence the solution. ■

Problem 5.6 Suggest an unbiased estimator for the population total under stratified sampling when *ppswr* sampling is used in all the L strata and also derive its variance.

Solution Let X_{hj} be the value of the size variable corresponding to the *j*th unit in the stratum *h*, $j = 1, 2, ..., N_h; h = 1, 2, ..., L$, X_h be the stratum total of the size variable corresponding to the stratum *h* and $P_{hj} = \frac{X_{hj}}{X_h}$. If p_{hj} is the *P*-value of the *j*th sampled in the stratum h, then an unbiased estimator of the population total is $\hat{Y}_{pt} = \sum_{h=1}^{L}\frac{1}{n_h}\sum_{j=1}^{n_h}\frac{y_{hj}}{p_{hj}}$. This estimator is constructed by using the fact that $\frac{1}{n_h}\sum_{j=1}^{n_h}\frac{y_{hj}}{p_{hj}}$ is unbiased for the stratum total Y_h of the study variable Y and hence by Theorem 5.1, $\hat{Y}_{pt} = \sum_{h=1}^{L}\frac{1}{n_h}\sum_{j=1}^{n_h}\frac{y_{hj}}{p_{hj}}$ is unbiased for the population total. The variance of the above estimator is

$$V(\hat{Y}_{pt}) = \sum_{h=1}^{L}\frac{1}{n_h}\sum_{j=1}^{n_h}\left\{\frac{Y_{hj}}{P_{hj}} - Y_h\right\}^2 P_{hj} \quad \text{(refer Theorem 4.2)}$$

Hence the solution. ■

Exercises

5.1 Derive the variance of the estimator considered in Problem 6.6 under proportional allocation. That is, the sample size is made proportional to stratum size X_h rather than the number of units in the stratum *h*.

5.2 A random sample of size n is selected from a population containing *N* units and the sample units are allocated *L* strata on the basis of information collected about them. Denoting by n_h the number of sample units failing in stratum h, derive the variance of $\sum_{h=1}^{L} \frac{N_h}{N} \overline{Y}_h$ (note that $\frac{N_h}{N}$ is known).

5.3 A population is divided into *L* strata, stratum *h* containing N_h units from which $n_h, h = 1, 2, ..., L$ are to be taken into the sample. The following procedure is used. One unit is selected with *pp* to *x* from the entire population. If the unit comes from the stratum *h*, a simple random sample of further $n_h - 1$ units is taken from the $N_h - 1$ units that remain. From the other strata simple random samples of specified sizes are taken. Show that under usual notations $\frac{\sum_{h=1}^{L} N_h \overline{y}_h}{\sum_{h=1}^{L} N_h \overline{x}_h}$ is an unbiased estimator of $\frac{Y}{X}$.

5.4 For the sampling scheme in which the population is split at random into substrata containing $N_i, i = 1, 2, ..., n$ units, and one unit is selected with *pp* to *x* from each substratum, suggest an unbiased estimator for the population total and derive its variance. (Compare this with Random group method described in Chapter 4).

Chapter 6

Use of Auxiliary Information

6.1 Introduction

So far we have seen many sampling-estimating strategies in which the knowledge of the variable under study, y, alone is directly used during the estimation stage. However in many situations the variable under study, y, will be closely related to an auxiliary variable x and information pertaining to it for all the units in the population is either readily available or can be easily collected. In such situations, it is desirable to consider estimators of the population total Y that use the data on x which are more efficient than the conventional ones. Two such methods are (i) ratio methods and (ii) regression methods. In the following sections we shall discuss "Ratio estimation".

6.2 Ratio Estimation

Let $\hat{Y}$ and $\hat{X}$ be unbiased for the population totals Y and X of the study and auxiliary variable respectively. The ratio estimator of the population total is defined as

$$\hat{Y}_R \doteq \frac{\hat{Y}}{\hat{X}} X \tag{6.1}$$

For example, if Y is the number of teak trees in a geographical region and x is its area in acres, the ratio $\frac{\hat{Y}}{\hat{X}}$ is an estimator of the number of teak trees per acre of a region in the population. The product of $\frac{\hat{Y}}{\hat{X}}$ with X, the total area in acres would provide an estimator of Y, the total number of teak trees in the population. The estimator proposed above is meant for any sampling design yielding unbiased estimators for the population totals Y and X. Let $P(s)$ be any sampling design. It may be noted that

$$E_P[\hat{Y}_R] = \sum_{s\in\Omega} \hat{Y}_R P(s) = \sum_{s\in\Omega}\left[\frac{\hat{Y}(s)}{\hat{X}(s)}\right] X P(s)$$

$$= \sum_{s\in\Omega}\left[\frac{\hat{Y}(s)}{\hat{X}(s)}\right] P(s)$$

Since the right hand side of the above expression is not equal to Y, the ratio estimator is biased for Y under the given sampling design.

The following theorem gives the approximate bias and mean square error of the ratio estimator.

Theorem 6.1 The approximate bias and mean square error of the ratio estimator are $B(\hat{Y}_R) = Y\left\{\left[\frac{V(\hat{X})}{X^2}\right] - \left[\frac{\text{cov}(\hat{X},\hat{Y})}{XY}\right]\right\}$

and $MSE(\hat{Y}_R) = Y^2\left\{\left[\frac{V(\hat{Y})}{Y^2}\right] + \left[\frac{V(\hat{X})}{X^2}\right] - 2\left[\frac{\text{cov}(\hat{X},\hat{Y})}{XY}\right]\right\}$

Proof Let $e_0 = \frac{\hat{Y}-Y}{Y}$ and $e_1 = \frac{\hat{X}-X}{X}$

It may be noted that (i) $E(e_0) = E\left[\frac{\hat{Y}-Y}{Y}\right] = 0$ (6.2)

$$\text{(ii) } E(e_1) = E\left[\frac{\hat{X}-X}{X}\right] = 0 \quad (6.3)$$

$$\text{(iii) } E(e_0^2) = E\left[\frac{\hat{Y}-Y}{Y}\right]^2 = \frac{V(\hat{Y})}{Y^2} \quad (6.4)$$

$$\text{(iv) } E(e_1^2) = E\left[\frac{\hat{X}-X}{X}\right]^2 = \frac{V(\hat{X})}{X^2} \quad (6.5)$$

$$\text{(v) } E(e_0 e_1) = E\left[\frac{(\hat{Y}-Y)(\hat{X}-X)}{YX}\right] = \frac{\text{cov}(\hat{Y},\hat{X})}{YX} \quad (6.6)$$

Assume that the sample size is large enough so that $|e_0|<1$ and $|e_1|<1$. This is equivalent to assuming that for all possible samples $0<\hat{X}<2X$ and $0<\hat{Y}<2Y$. Since $\hat{Y} = Y(1+e_0)$ and $\hat{X} = X(1+e_1)$, the estimator $\hat{Y}_R = \frac{\hat{Y}}{\hat{X}}X$ can be written as

$$\begin{aligned}\hat{Y}_R &= Y(1+e_0)(1+e_1)^{-1} \\ &= Y(1+e_0)(1-e_1+e_1^2-\ldots) \\ &= Y(1+e_0-e_1+e_1^2-e_0e_1+\ldots)\end{aligned}$$

Using (6.2) and (6.3) and ignoring terms of degree greater than two we get

$$\begin{aligned}E[\hat{Y}_R - Y] &\cong YE[e_1^2 - e_0e_1] \\ &= Y\left\{\frac{V(\hat{X})}{X^2} - \frac{\text{cov}(\hat{Y},\hat{X})}{YX}\right\} \quad \text{(using (6.4),(6.5) and (6.6))}\end{aligned}$$

Proceeding as above we get (ignoring terms of degree greater than two)

$$E[\hat{Y}_R - Y]^2 \cong Y^2 E[e_0^2 + e_1^2 - 2e_0 e_1]$$
$$= Y^2 \left\{ \frac{V(\hat{Y})}{Y^2} + \frac{V(\hat{X})}{X^2} - 2\frac{\text{cov}(\hat{Y}, \hat{X})}{YX} \right\}$$

Hence the proof. ■

Corollary 6.1 Under simple random sampling,

$$\text{(i)} \quad \hat{Y}_R = \frac{\sum_{i \in s} Y_i}{\sum_{i \in s} X_i} X \quad \text{(ii)} \quad B(\hat{Y}_R) = \frac{N^2(N-n)}{Nn} \left\{ \frac{S_x^2}{X^2} - \frac{S_{xy}}{XY} \right\}$$

$$\text{(iii)} \; MSE(\hat{Y}_R) = \frac{N^2(N-n)}{Nn} \left\{ \frac{S_y^2}{Y^2} + \frac{S_x^2}{X^2} - 2\frac{S_{xy}}{XY} \right\}$$

Proof We know that under simple random sampling $V(\hat{Y}) = \frac{N^2(N-n)}{Nn} S_y^2$

$V(\hat{X}) = \frac{N^2(N-n)}{Nn} S_x^2$ and $\text{cov}(\hat{Y}, \hat{X}) = \frac{N^2(N-n)}{Nn} S_{xy}$. Substituting these expression in the results available in Theorem 6.1, we get the required expressions. ■

The following theorem gives the condition under which the ratio estimator will be more efficient than the conventional expansion estimator.

Theorem 6.2 The ratio estimator $\hat{Y}_R = \frac{\hat{Y}}{\hat{X}} X$ is more efficient than the expansion estimator $\hat{Y}$ if $\rho > \frac{1}{2}\frac{C_x}{C_y}$ where $C_y = \frac{S_y}{\overline{\overline{Y}}}, C_x = \frac{S_x}{\overline{\overline{X}}}$ and ρ is the coefficient of correlation.

Proof $V(\hat{Y}) > MSE(\hat{Y}_R)$

$$\Rightarrow N^2 \frac{N-n}{NN} S_y^2 > N^2 \frac{N-n}{Nn} Y^2 \left\{ \frac{S_y^2}{Y^2} + \frac{S_x^2}{X^2} - 2\frac{S_{xy}}{XY} \right\}$$

$$S_y^2 > \left\{ S_y^2 + \frac{Y^2}{X^2} S_x^2 - 2\frac{Y}{X} S_{xy} \right\}$$

$$2S_{xy} > \frac{Y}{X} S_x^2$$

$$S_{xy} > \frac{1}{2}\frac{Y}{X}\frac{S_x}{S_y}$$

$$\rho > \frac{1}{2}\frac{C_x}{C_y}$$

Hence the proof. ■

Estimated mean square error under simple random sampling

Note that

$$\sum_{i=1}^{N}[Y_i - RX_i]^2 = \sum_{i=1}^{N}[Y_i - \overline{Y} + \overline{Y} - RX_i]^2$$

$$= \sum_{i=1}^{N}[Y_i - \overline{Y} + R\overline{X} - RX_i]^2 \quad (\text{since } R = \frac{\overline{Y}}{\overline{X}})$$

$$= \sum_{i=1}^{N}[Y_i - \overline{Y}]^2 + R^2\sum_{i=1}^{N}[X_i - \overline{X}]^2 - 2R\sum_{i=1}^{N}[Y_i - \overline{Y}][X_i - \overline{X}]$$

Dividing both the sides by $(N-1)$ we get

$$\frac{1}{(N-1)}\sum_{i=1}^{N}[Y_i - RX_i]^2 = S_y^2 + R^2 S_x^2 - 2RS_{xy} \tag{6.10}$$

Substituting this in the expression for the mean square error, we get an equivalent expression as

$$MSE(\hat{Y}_R) = \frac{N^2(N-n)}{Nn}\frac{1}{(N-1)}\sum_{i=1}^{N}[Y_i - RX_i]^2 \quad \text{(replacing (6.10))}$$

Therefore a reasonable estimate for the mean square error of the ratio estimate is

$$\frac{N^2(N-n)}{Nn}\frac{1}{(n-1)}\sum_{i \in s}[Y_i - \hat{R}X_i]^2 \text{ where } \hat{R} = \frac{\sum_{i \in s} Y_i}{\sum_{i \in s} X_i}$$

The ratio estimator considered in this section is not unbiased for the population total (mean). In the following section of this chapter, few ratio type unbiased estimators meant for simple random sampling are presented.

6.3 Unbiased Ratio Type Estimators

Already we have seen that under simple random sampling, the ratio estimator takes the form $\left[\frac{\sum_{i \in s} Y_i}{\sum_{i \in s} X_i}\right] X$. As an alternative to this, it is reasonable to take $\hat{Y}_{RO} = \left[\frac{N}{n}\right]\sum_{i \in s}\left[\frac{Y_i}{X_i}\right]\overline{X}$ as estimator of the population total. Like the ratio estimator, the above estimator is also biased for the population total. The following theorem gives an expression for the bias of $\hat{Y}_{RO}$.

Theorem 6.3 The bias of the estimator $\hat{Y}_{RO}$ is $B(\hat{Y}_{R_O}) = -[N-1]S_{zx}$

where $S_{zx} = \frac{1}{N-1}\sum_{i-1}^{N}[Z_i - \overline{Z}][X_i - \overline{X}], Z_i = \frac{Y_i}{X_i}$

Proof Taking $Z_i = \frac{Y_i}{X_i}$, $i = 1, 2, ..., N$, the estimator $\hat{Y}_{RO}$ can be written as

$$\hat{Y}_{RO} = \frac{N}{n}\sum_{i\in s} Z_i \overline{X}$$

$$= \hat{Z}\overline{X} \text{ where } \hat{Z} = \frac{N}{n}\sum_{i\in s} Z_i$$

The bias of $\hat{Y}_{RO}$ is $B(\hat{Y}_{R_O}) = E[\hat{Y}_{R_O} - Y]$

$$= E(\hat{Z}\overline{X}) - Y$$

$$= Z\overline{X} - Y \text{ where } Z = \sum_{i=1}^{N} Z_i \tag{6.11}$$

We know that $\operatorname{cov}(\hat{Z}, \hat{X}) = \frac{N^2(N-n)}{Nn}\frac{1}{N-1}\sum_{i=1}^{N}[Z_i - \overline{Z}][X_i - \overline{X}]$

$$= \frac{N^2(N-n)}{Nn}\frac{1}{N-1}\left\{\sum_{i=1}^{N} Z_i X_i - \overline{Z}\,\overline{X}\right\}$$

$$= \frac{N^2(N-n)}{Nn}\frac{1}{N-1}\left\{N\overline{Y} - N\,\overline{Z}\,\overline{X}\right\} \text{ using } Z_i = \frac{Y_i}{X_i}$$

$$= \frac{N^2(N-n)}{Nn}\frac{1}{N-1} B(\hat{Y}_{R_0}) \text{ (using (6.11))}$$

Therefore the bias of the estimator $\hat{Y}_{R_O}$ is

$$B(\hat{Y}_{R_O}) = -\frac{Nn(N-1)}{N^2(N-n)}\operatorname{cov}(\hat{Z}, \hat{X}) \tag{6.12}$$

We know that under simple random sampling, $\operatorname{cov}(\hat{Y}, \hat{X}) = \frac{N^2(N-n)}{Nn} S_{xy}$

Making use of this result in (6.12) we get the required expression. ■

The above theorem helps us to get an unbiased estimator for the population total as shown below.

We have seen in Chapter 2 that $s_{xy} = \frac{1}{n-1}\sum_{i\in s}(X_i - \hat{\overline{X}})(Y_i - \hat{\overline{Y}})$ is unbiased for

$S_{xy} = \frac{1}{N-1}\sum_{i=1}^{N}(X_i - \overline{X})(Y_i - \overline{Y})$. Therefore an unbiased estimator for the bias given in Theorem 6.3 is

$$B(\hat{Y}_{R0}) = -(N-1)S_{zx}$$

$$= \frac{-(N-1)}{n-1}\sum_{i\in s}(Z_i - \hat{\bar{Z}})(X_i - \hat{\bar{X}})$$

$$= \frac{-(N-1)}{n-1}\left\{\sum_{i\in s} Z_i X_i - n\hat{\bar{Z}}\,\hat{\bar{X}})\right\}$$

$$= \frac{-(N-1)}{n-1}\left\{\sum_{i\in s} Y_i - n\hat{\bar{Z}}\,\hat{\bar{X}})\right\} \text{ (using } Z_i = \frac{Y_i}{X_i})$$

$$= \frac{-n(N-1)}{n-1}[\hat{\bar{Y}} - \hat{\bar{Z}}\,\hat{\bar{X}}]$$

It may be observed that, if b is an unbiased estimator of the bias of the estimator T (which is meant for estimating the parameter θ) then T-b is unbiased for the parameter θ. Therefore $\hat{Y}_{R0} - B(\hat{Y}_{R0})$ is an unbiased estimator of the population total. That is, $\hat{Y}_{R0} + \frac{n(N-1)}{n-1}[\hat{\bar{Y}} - \hat{\bar{Z}}\,\hat{\bar{X}}]$ is unbiased for the population total Y.

Thus we have obtained an exactly unbiased ratio-type estimator by considering the mean of the ratios of Y_i to X_i (instead of the ratio of sum of Y_i to sum of X_i) to form the estimator and correcting for the bias. The above estimator is due to Hartley and Rao (1954). In the following section another corrected estimator is presented.

6.4 Almost Unbiased Ratio Estimator

Suppose a sample of size n is drawn in the form of m independent sub-samples of the same size, selected according to the same sampling design and $\hat{Y}_i$ and $\hat{X}_i$, $i = 1, 2, ..., m$ are unbiased estimates of the population totals Y and X based on the m subsamples. The following two estimates can be considered for Y:

$$\hat{Y}_1 = \frac{\hat{Y}}{\hat{X}} X \tag{6.13}$$

where $\hat{Y} = \frac{1}{m}\sum_{i=1}^{m} Y_i$ and $\hat{X} = \frac{1}{m}\sum_{i=1}^{m} X_i$

and

$$\hat{Y}_m = \frac{1}{m}\sum_{i=1}^{m}\frac{\hat{Y}_i}{\hat{X}_i} X$$

$$= \frac{X}{m}\sum_{i=1}^{m} r_i \tag{6.14}$$

where $r_i = \frac{\hat{Y}_i}{\hat{X}_i}$.

Under the usual assumptions (stated in the proof of Theorem 6.1), the bias of the estimator $\hat{Y}_1$ is

$$B_1 = Y[RV(\hat{X}) - \text{cov}(\hat{X}, \hat{Y})] \quad \text{(by Theorem 6.1)}$$

$$= Y\left[RV\left\{\frac{1}{m}\sum_{i=1}^{m}\hat{X}_i\right\} - \text{cov}\left\{\frac{1}{m}\sum_{i=1}^{m}\hat{X}_i, \frac{1}{m}\sum_{i=1}^{m}\hat{Y}_i\right\}\right]$$

$$= \frac{1}{m^2}Y\sum_{i=1}^{m}\left[RV(\hat{X}_i) - \text{cov}(\hat{X}_i, \hat{Y}_i)\right]$$

$$= \frac{1}{m^2}\sum_{i=1}^{m}B(r_i)$$

where $B(r_i) = Y[RV(\hat{X}_i) - \text{cov}(\hat{X}_i, \hat{Y}_i)]$ (6.15)

and the bias of the estimator $\hat{Y}_m$ is

$$B_m = B(\hat{Y}_m)$$

$$= \frac{1}{m}\sum_{i=1}^{m}B(r_i) \tag{6.16}$$

Comparing (6.15) and (6.16) we get $mB_1 = B_m$ (6.17)

This shows that the bias of the estimator $\hat{Y}_m$ is m times that of $\hat{Y}_1$. Further it can be seen that $B_m - B_1 = E[\hat{Y}_m - Y] - E[\hat{Y}_1 - Y]$

$$= E[\hat{Y}_m - \hat{Y}_1]$$

Therefore $E[\hat{Y}_m - \hat{Y}_1] = (m-1)B_1$.

Hence $\frac{[\hat{Y}_m - \hat{Y}_1]}{m-1}$ is an unbiased estimator of B_1.

Thus after correcting the estimator $\hat{Y}_1$ for its bias, we get an unbiased estimator for the population total

$$\hat{Y}_{AU} = \hat{Y}_1 - \frac{[\hat{Y}_m - \hat{Y}_1]}{m-1} = \frac{[m\hat{Y}_1 - \hat{Y}_m]}{m-1} \tag{6.18}$$

Since the estimator given above is obtained by correcting only the approximate bias (not the exact bias), it is known as "Almost Unbiased Ratio-Type Estimator".

.6.5 Jackknife Ratio Estimator

As in the previous section, here also it is assumed that a simple random sample of size n is selected in the form of m independent subsamples of k units each. Let $\hat{Y}_1 = \frac{\hat{Y}}{\hat{X}} X$ where $\hat{Y} = \frac{1}{m}\sum_{i=1}^{m} \hat{Y}_i$ and $\hat{X} = \frac{1}{m}\sum_{i=1}^{m} \hat{X}_i$. Further denote by

$\hat{Y}_{Ri} = \frac{\hat{Y}^{(i)}}{\hat{X}^{(i)}} X$ where $\hat{X}^{(i)}$ and $\hat{Y}^{(i)}$ are unbiased estimators of X and Y

obtained after omitting the ith subsample. That is, $\hat{Y}_{Ri}$ is the ratio estimate computed after omitting the ith subsample. Combining $\hat{Y}_1$ and $\hat{Y}_{Ri}$, Quenouille (1956) suggested the estimator

$$\hat{Y}_Q = m\hat{Y}_1 - \frac{m-1}{m}\sum_{i=1}^{m} \hat{Y}_{Ri} \tag{6.19}$$

The above estimator is popularly known as Jackknife ratio estimator.

In the following theorem it is proved that the above estimator is also approximately unbiased.

Theorem 6.4 The estimator $\hat{Y}_Q = m\hat{Y}_1 - \frac{m-1}{m}\sum_{i=1}^{m} \hat{Y}_{Ri}$ is approximately unbiased.

Proof The bias of the estimator $\hat{Y}_Q = m\hat{Y}_1 - \frac{m-1}{m}\sum_{i=1}^{m} \hat{Y}_{Ri}$ is

$$\begin{aligned} B(\hat{Y}_Q) &= E(\hat{Y}_Q) - Y \\ &= E\left[m\hat{Y}_1 - \frac{m-1}{m}\sum_{i=1}^{m} \hat{Y}_{Ri} \right] - Y \\ &= E\left[m(\hat{Y}_1 - Y) - \frac{m-1}{m}\sum_{i=1}^{m} \hat{Y}_{Ri} \right] + (m-1)Y \\ &= E\left[m(\hat{Y}_1 - Y) - \frac{m-1}{m}\sum_{i=1}^{m} (\hat{Y}_{Ri} - Y) \right] \\ &= mB(\hat{Y}_1) - \frac{m-1}{m}\sum_{i=1}^{m} B(\hat{Y}_{Ri}) \end{aligned} \tag{6.20}$$

We have seen in the previous section that the bias of the estimator $\hat{Y}_1$ is

$$B(\hat{Y}_1) = \frac{1}{m}\sum_{i=1}^{m} B(\hat{r}_i)$$

where $B(\hat{r}_i)$ is as defined in (6.15).

Since the subsamples are drawn independently and they are of the same size, $B(\hat{r}_i) = B_0$ (constant) for $i = 1, 2, \ldots, m$.

$$\text{Hence } B(\hat{Y}_1) = \frac{1}{m^2} m B_0$$

$$= \frac{B_0}{m} \tag{6.21}$$

Note that for each i, $\frac{1}{m-1}\sum_{\substack{j=1\\ \neq i}}^{m} \hat{X}_j$ and $\frac{1}{m-1}\sum_{\substack{j=1\\ \neq i}}^{m} \hat{Y}_j$ are unbiased for the population totals X and Y respectively. Therefore by Theorem 6.1, the bias of the estimator $\hat{Y}_{Ri}$ is

$$B(\hat{Y}_{Ri}) = Y[RV(\hat{X}^{(i)}) - \text{cov}(\hat{X}^{(i)}, \hat{Y}^{(i)})]$$

$$= Y\left\{RV\left[\frac{1}{m-1}\sum_{\substack{j=1\\ \neq i}}^{m} \hat{X}_j\right] - \text{cov}\left(\left[\frac{1}{m-1}\sum_{\substack{j=1\\ \neq i}}^{m} \hat{X}_j\right], \left[\frac{1}{m-1}\sum_{\substack{j=1\\ \neq i}}^{m} \hat{Y}_j\right]\right)\right\}$$

$$= \frac{1}{(m-1)^2} Y\left(\sum_{\substack{j=1\\ \neq i}}^{m} [RV(\hat{X}_j) - \text{cov}(\hat{X}_j, \hat{Y}_j)]\right)$$

$$= \frac{1}{(m-1)^2}\left[\sum_{\substack{j=1\\ \neq i}}^{m} B(\hat{r}_j)\right] = \frac{(m-1)B_0}{(m-1)^2}$$

$$= \frac{B_0}{(m-1)} \tag{6.22}$$

Substituting (6.21) and (6.22) in (6.20) we get

$$B(\hat{Y}_Q) = m\frac{B_0}{m} - \frac{m-1}{m} m \frac{B_0}{m-1}$$

$$= 0$$

Therefore the Jackknife estimator given in (6.19) is approximately unbiased. It is pertinent to note that this estimator is not an exactly unbiased estimator. ■

6.6 Bound for Bias

In the last three sections, we have seen unbiased ratio-type estimators. In this section, an upper bound is presented for the bias of the ratio estimator.

We know that the bias of the ratio estimator $\hat{Y}_R$ is

$$B[\hat{Y}_R] = E[\hat{Y}_R] - Y$$

$$\text{and } \quad \text{cov}(\hat{Y}_R, \hat{X}) = E[\hat{Y}_R X] - E[\hat{Y}_R]E[\hat{X}]$$

$$= E\left[\frac{\hat{Y}}{\hat{X}} X\right] - E[\hat{Y}_R]E[\hat{X}]$$

$$= X\, E[\hat{Y}] - E[\hat{Y}_R]X$$

$$= XY - E[\hat{Y}_R]X$$

$$= -XB[\hat{Y}_R] \quad \text{(using (6.22))}$$

Therefore $cor(\hat{Y}_R, \hat{X}) \le SD(\hat{Y}_R)SD(\hat{X}) = -X\, B[\hat{Y}_R]$

$$\Rightarrow SD(\hat{Y}_R)SD(\hat{X}) \le X\left|B[\hat{Y}_R]\right|$$

Hence $$\frac{\left|B[\hat{Y}_R]\right|}{SD(\hat{Y}_R)} \le \frac{SD(\hat{X})}{X}$$

The above bound is due to **Hartley and Ross(1954).**

6.7 Product Estimation

We have proved under simple random sampling, the ratio estimator is more precise than the expansion estimator when the variables x and y have high positive correlation. In fact, it is not difficult to see under any sampling design, $\hat{Y}_R$ is more efficient than $\hat{Y}$ if $\rho(\hat{X}, \hat{Y}) > \frac{1}{2}\left[\frac{C(\hat{X})}{C(\hat{Y})}\right]$ where $C(\hat{Y}) = \frac{SD(\hat{Y})}{Y}$ and $C(\hat{X}) = \frac{SD(\hat{X})}{X}$. This shows that if the correlation between x and y is negative, the ratio estimator will not be precise than the conventional estimator. For such situations, Murthy(1964) suggested another method of estimation, which is expected to be more efficient than $\hat{Y}$ in situations, where $\hat{Y}_R$ turns out to be less efficient than $\hat{Y}$. In this method, termed "Product method of estimation, the population total is estimated by using the estimator

$$\hat{Y}_P = \frac{\hat{Y}}{\hat{X}}\hat{X} \qquad (6.24)$$

Since the estimator uses the product $\hat{Y}\hat{X}$ rather than the ratio $\frac{\hat{Y}}{\hat{X}}$, it is known as product estimator.

The following theorem gives the exact bias and approximate mean square error of the ratio estimator.

Theorem 6.5 The exact bias and the approximate mean square error of the product estimator are given by

$$B[\hat{Y}_P] = \frac{\text{cov}(\hat{X}, \hat{Y})}{X}$$

and $$MSE[\hat{Y}_P] = V(\hat{Y}) + 2R\frac{\text{cov}(\hat{X},\hat{Y})}{X} + R^2V(\hat{X})$$

Proof Using the notations and assumptions introduced in Theorem 6.1, the estimator $\hat{Y}_P$ can be written as

$$\hat{Y}_P = Y(1+e_0)(1+e_1)$$
$$= Y(1+e_0+e_1+e_0+e_1)$$

Therefore $\hat{Y}_P - Y = Y[e_0+e_1+e_0+e_1]$ (6.25)

Taking expectation on both the sides of (6.25) we get

$$\hat{Y}_P - Y = YE[e_0e_1] \quad (\text{since } E(e_0) = E(e_1) = 0)$$
$$= Y\left[\frac{\text{cov}(\hat{Y},\hat{X})}{YX}\right]$$
$$\doteq \frac{\text{cov}(\hat{Y},\hat{X})}{X} \tag{6.26}$$

Squaring and taking expectation on both the sides of (6.25) and ignoring terms of degree greater than two, we get the approximate mean square error

$$E[\hat{Y}_P - Y]^2 = Y^2E[e_0^2+e_1^2+2e_0e_1]$$
$$= Y^2\left\{\left[\frac{V(\hat{Y})}{Y^2}\right]+\left[\frac{V(\hat{X})}{X^2}\right]+2\left[\frac{\text{cov}(\hat{Y},\hat{X})}{YX}\right]\right\}$$
$$= V(\hat{Y}) + R^2V(\hat{X}) + 2R\,\text{cov}(\hat{Y},\hat{X}) \tag{6.27}$$

Hence the proof. ■

The following theorem gives the condition under which the estimator $\hat{Y}_P$ will be more efficient than $\hat{Y}$.

Theorem 6.6 The product estimator $\hat{Y}_P$ is more efficient than $\hat{Y}$ if

$$\rho(\hat{X},\hat{Y}) < -\frac{1}{2}\left[\frac{C(X)}{C(Y)}\right]$$

Proof Left as an exercise.

Theorem 6.7 Under simple random sampling $\hat{Y}_P + \left[\frac{N^2(N-n)}{Nn}\right]\left(\frac{s_{xy}}{\overline{X}}\right)$ is unbiased for the population total.

Proof We know that under simple random sampling $\text{cov}(\hat{Y},\hat{X}) = \left[\frac{N^2(N-n)}{Nn}\right]S_{xy}$. Therefore the true bias of the product estimator under simple random sampling is $-\left[\frac{N^2(N-n)}{Nn}\right]\frac{S_{xy}}{X}$. Since s_{xy} is unbiased for S_{xy} under simple random sampling, an unbiased estimator of the bias of the product estimator is $-\left[\frac{N^2(N-n)}{Nn}\right]\frac{s_{xy}}{X}$. Therefore after adjusting the product estimator for its bias we get $\hat{Y}_P+\left[\frac{N^2(N-n)}{Nn}\right]\left(\frac{s_{xy}}{\overline{X}}\right)$ as unbiased estimator of the population total. It may be noted that the above estimator is not an exact unbiased estimator. ■

6.8 Two Phase Sampling

The ratio and product estimators introduced in this chapter assume the knowledge of the population total X of the auxiliary variable x. However there are some situations where the population total of the auxiliary variable will not be known in advance. In such cases, two-phase sampling can be used for getting ratio or product estimator. In two phase sampling, a sample of size n' is selected initially by using a suitable sampling design and the population total X is estimated and then a sample of size n is selected to estimate the population totals of the study and auxiliary variables. The second phase sample can be either a subsample of the first phase sample or it can be directly drawn from the given population. The sampling designs used in the first and second phases need not be the same. Depending on the situation, different sampling designs can also be used. Generally, two-phase sampling is recommended only when the cost of conducting first phase survey is more economical when compared to that of the second phase.

Let $\hat{X}_D$ be an unbiased estimator of X based on the first phase sample and $\hat{X},\hat{Y}$ be unbiased estimators of X, Y based on the second phase sample. Then the ratio and product estimators based on two-phase sampling are

$$\hat{Y}_{RD}=\frac{\hat{Y}}{\hat{X}}\hat{X}_D \tag{6.29}$$

and

$$\hat{Y}_{PD}=\frac{\hat{Y}\hat{X}}{X} \tag{6.30}$$

The following theorems give the approximate bias and mean square error of the ratio and product estimator under different cases of two-phase sampling.

Theorem 6.8 (i) When the samples are drawn independently in the two phases of sampling the approximate bias of the ratio estimator is

$$B[\hat{Y}_{RD}]=Y\left\{\left[\frac{V(\hat{X})}{X^2}\right]-\left[\frac{\text{cov}(\hat{X},\hat{Y})}{XY}\right]\right\}$$

(ii) When the second phase sample is a subsample of the first phase sample, the approximate bias of the ratio estimator is

$$B[\hat{Y}_{RD}] = Y\left\{\left[\frac{V(\hat{X})}{X^2}\right] - \left[\frac{\text{cov}(\hat{X},\hat{Y})}{XY}\right] - \left[\frac{\text{cov}(\hat{X},X_D)}{X^2}\right] + \left[\frac{\text{cov}(\hat{Y},\hat{X}_D)}{XY}\right]\right\}$$

Proof When the samples are drawn independently in the two phases of sampling

$$\text{cov}(\hat{X},\hat{X}_D) = 0 \tag{6.31}$$

$$\text{cov}(\hat{Y},\hat{X}_D) = 0 \tag{6.32}$$

Let $e_0 = \dfrac{\hat{Y}-Y}{Y}$, $e_1 = \dfrac{\hat{X}-X}{X}$ and $e_d = \dfrac{\hat{X}_d - X}{X}$

It may be noted that

(i) $E(e_0) = E\left[\dfrac{\hat{Y}-Y}{Y}\right] = 0$, (ii) $E(e_1) = E\left[\dfrac{\hat{X}-X}{X}\right] = 0$

(iii) $E(e_d) = E\left[\dfrac{\hat{X}_d - X}{X}\right] = 0$ (iv) $E(e_0^2) = E\left[\dfrac{\hat{Y}-Y}{Y}\right]^2 = \dfrac{V(\hat{Y})}{Y^2}$

(v) $E(e_1^2) = E\left[\dfrac{\hat{X}-X}{X}\right]^2 = \dfrac{V(\hat{X})}{X^2}$ (vi) $E(e_d^2) = E\left[\dfrac{\hat{X}_D - X}{X}\right]^2 = \dfrac{V(\hat{X}_D)}{X^2}$

(vii) $E(e_0 e_1) = E\left[\dfrac{(\hat{Y}-Y)(\hat{X}-X)}{YX}\right] = \dfrac{\text{cov}(\hat{Y},\hat{X})}{YX}$

(vii) $E(e_0 e_d) = E\left[\dfrac{(\hat{Y}-Y)(\hat{X}_D - X)}{YX}\right] = \dfrac{\text{cov}(\hat{Y},\hat{X}_D)}{YX}$

(ix) $E(e_1 e_d) = E\left[\dfrac{(\hat{X}-X)(\hat{X}_D - X)}{YX}\right] = \dfrac{\text{cov}(\hat{X},\hat{X}_D)}{YX}$

The ratio estimator can be expressed in terms of e_0, e_1, e_d as follows :

$$\begin{aligned}
\hat{Y}_R &= Y(1+e_0)(1+e_1)^{-1}(1+e_d) \\
&= Y(1+e_0)(1+e_d)(1-e_1+e_1^2-\ldots\ldots..) \\
&= Y(1+e_0+e_d+e_0e_d)(1-e_1+e_1^2-e_1^3+\ldots\ldots..) \\
&= Y(1-e_1+e_1^2+e_0-e_0e_1+e_d-e_1e_d+e_0e_d)
\end{aligned}$$

(ignoring terms of degree greater than two)

This implies $\hat{Y}_{RD} - Y = Y(e_0 - e_1 + e_d + e_1^2 - e_0e_1 - e_1e_d + e_0e_d)$ (6.33)

Taking expectations on both the sides of (6.33) and using expressions above we get when the samples are drawn independently drawn, the approximate bias of the ratio estimator as $B[\hat{Y}_{RD}] = Y\left\{\left[\dfrac{V(\hat{X})}{X^2}\right] - \left[\dfrac{\text{cov}(\hat{X},\hat{Y})}{XY}\right]\right\}$

The bias of the ratio estimator when the second phase sample is a subsample of the first phase sample can be obtained by taking expectations on both the sides of (6.33) as

$$B[\hat{Y}_{RD}] = Y\left\{\left[\frac{V(\hat{X})}{X^2}\right] - \left[\frac{\text{cov}(\hat{X},\hat{Y})}{XY}\right] - \left[\frac{\text{cov}(\hat{X},X_D)}{X^2}\right] + \left[\frac{\text{cov}(\hat{Y},\hat{X}_D)}{XY}\right]\right\}$$

Hence the proof. ■

Theorem 6.9 (i) When the samples are drawn independently in the two phases of sampling, the approximate mean square error of the ratio estimator is

$$MSE[\hat{Y}_{RD}] = Y^2\left\{\left[\frac{V(\hat{Y})}{Y^2}\right] + \left[\frac{V(\hat{X})}{X^2}\right] + \left[\frac{V(\hat{X}_D)}{X^2}\right] - 2\left[\frac{\text{cov}(\hat{X},\hat{Y})}{XY}\right]\right\}$$

(ii) When the second phase sample is a subsample of the first phase sample, the approximate mean square error of the ratio estimator is

$$MSE[\hat{Y}_{RD}] = Y^2\left\{\left[\frac{V(\hat{X})}{X^2}\right] + \left[\frac{V(\hat{X})}{X^2}\right] + \left[\frac{V(\hat{X})}{X^2}\right] - 2\left[\frac{\text{cov}(\hat{X},\hat{Y})}{XY}\right]\right.$$
$$\left. - 2\left[\frac{\text{cov}(\hat{X},X_D)}{X^2}\right] + 2\left[\frac{\text{cov}(\hat{Y},\hat{X}_D)}{XY}\right]\right\}$$

Proof of this theorem is left as exercise.

Theorem 6.10 (i) When the samples are drawn independently, the approximate bias of the product estimator in two phase sampling is

$$B[\hat{Y}_{PD}] = Y\left\{\left[\frac{V(\hat{X})}{X^2}\right] + \left[\frac{\text{cov}(\hat{X},\hat{Y})}{XY}\right]\right\}$$

(ii) When the second phase sample is a subsample of the first phase sample, the approximate bias of the product estimator is

$$B[\hat{Y}_{PD}] = Y\left\{\left[\frac{V(\hat{X}_D)}{X^2}\right] + \left[\frac{\text{cov}(\hat{X},\hat{Y})}{XY}\right] - \left[\frac{\text{cov}(\hat{X},X_D)}{X^2}\right] - \left[\frac{\text{cov}(\hat{Y},\hat{X}_D)}{XY}\right]\right\}$$

Proof: Proof of this theorem is left as exercise.

Theorem 6.11 (i) When the samples are drawn independently in the two phases of sampling, he approximate mean square error of the product estimator is

$$MSE[\hat{Y}_{PD}] = Y^2\left\{\left[\frac{V(\hat{Y})}{Y^2}\right] + \left[\frac{V(\hat{X})}{X^2}\right] + \left[\frac{V(\hat{X}_D)}{X^2}\right] + 2\left[\frac{\text{cov}(\hat{X},\hat{Y})}{XY}\right]\right\}$$

(ii) When the second phase sample is a subsample of the first phase sample, the approximate mean square error of the product estimator is given by

$$MSE[\hat{Y}_{PD}] = Y^2\left\{ \frac{V(\hat{Y})}{Y^2} + \frac{V(\hat{X})}{X^2} + \frac{V(\hat{X}_D)}{X^2} + \right.$$

$$\left. +2\left[\frac{\text{cov}(\hat{X},\hat{Y})}{XY} - \frac{\text{cov}(\hat{X},X_D)}{X^2} + \frac{\text{cov}(\hat{Y},\hat{X}_D)}{XY}\right]\right\}$$

Proof of this theorem is left as exercise.

The theorems stated above are quite general in nature and they are applicable for any sampling design. Now we shall develop the approximate bias and mean square error of the ratio estimator under the two cases of two-phase sampling when simple random sampling is used in the two-phases of sampling. Towards this we observe the following :

Let s' and s be the samples obtained in the two-phases of sampling, where in the first phase a large sample of size n' is drawn to estimate the population total X and in the second phase a sample of size n is drawn to estimate the population totals X and Y both by using simple random sampling. Here n is assumed to be small when compared to n'.

Let $\hat{X}_D = \frac{N}{n'}\sum_{i \in s'} X_i$, $\hat{X} = \frac{N}{n}\sum_{i \in s} X_i$ and $\hat{Y} = \frac{N}{n}\sum_{i \in s} X_i$

Note that when s is a subsample of s',

$$\text{(i) } E(\hat{X}_D) = \sum_{i=1}^{N} X_i \tag{6.36}$$

$$\text{(ii) } E(\hat{X}) = E_I E_{II}[\hat{X} \mid s'] = E_I E_{II}\left[\frac{N}{n}\sum_{i \in s} X_i \mid s'\right]$$

$$= E_I\left[N^2(N-n')\sum_{i \in s} X_i\left(\frac{n'}{n}\right)\right] = X \tag{6.37}$$

$$\text{(iii) } V(\hat{X}_d) = \left[N^2\frac{(N-n)}{Nn}\right]S_x^2 \tag{6.38}$$

$$\text{(iv) } V(\hat{X}) = E_I V_{II}[\hat{X} \mid s'] + V_I E_{II}[\hat{X} \mid s']$$

$$= E_I\left[N^2\frac{(n'-n)}{n'n}\right]s_x^2 + V_I(\hat{X}_d)$$

$$= \left[N^2\frac{(N-n)}{Nn}\right]S_x^2 \tag{6.39}$$

$$\text{(v) } V(\hat{Y}_d) = \left[N^2\frac{(N-n)}{Nn}\right]S_y^2 \tag{6.40}$$

(vi) $\text{cov}(\hat{X}, \hat{X}_d) = E[\hat{X}\hat{X}_d] - E[\hat{X}]E[\hat{X}_d]$

$$= E_I E_{II}[\hat{X}\hat{X}_d \mid s'] - X^2$$

$$\doteq E_I[\hat{X}_d \hat{X}_d] - X^2$$

$$= \left[N^2 \frac{(N-n)}{Nn}\right] S_x^2 \qquad (6.41)$$

(vii) $$\text{cov}(\hat{Y}, \hat{X}_d) = \left[N^2 \frac{(N-n)}{Nn}\right] S_{xy} \qquad (6.42)$$

(viii) $\text{cov}(\hat{Y}, \hat{X}) = E[\hat{X}\hat{Y}] - XY$

$$= E_I E_{II}[\hat{X}\hat{Y} \mid s'] - XY$$

$$= E_I\left[N^2 \frac{(n'-n)}{n'n} s'_{xy} + Y_d X_d\right] - XY$$

$$= \left[N^2 \frac{(n'-n)}{n'n}\right] S_{xy} + E_I\left[X_d Y_d\right] - XY$$

$$= \left[N^2 \frac{(n'-n)}{n'n}\right] S_{xy} + \text{cov}(X_d, Y_d)$$

$$= \left[N^2 \frac{(n'-n)}{n'n}\right] S_{xy} + \left[N^2 \frac{(N-n')}{Nn'}\right] S_{xy}$$

$$= \left[N^2 \frac{(N-n)}{Nn}\right] S_{xy} \qquad (6.43)$$

Here $\hat{Y}_d$ and s'_{xy} are the analogues of $\hat{Y}$ and s_{xy} respectively based on the sample s'.

When the samples are drawn independently, the results derived in simple random sampling can be used directly without any difficulty.

The following theorem gives the approximate bias and mean square error of the ratio estimator when simple random sampling is used in both the phases.

Theorem 6.12 When simple random sampling is used in both the phases of sampling and the samples are drawn independently

$$B[\hat{Y}_{RD}] = \left[N^2 \frac{(N-n)}{Nn}\right] Y\left[C_{xx} - C_{xy}\right]$$

and

$$MSE[\hat{Y}_{RD}] = \left[N^2 \frac{(N-n)}{Nn}\right] Y^2\left[C_{yy} + C_{xx} - 2C_{xy}\right] + \left[N^2 \frac{(N-n')}{Nn'}\right] Y^2 C_{xx}$$

where $C_{xx} = \dfrac{S_x^2}{X^2}$, $C_{yy} = \dfrac{S_y^2}{Y^2}$ and $C_{xy} = \dfrac{S_{xy}}{XY}$.

Theorem 6.13 When simple random sampling is used in both the phases of sampling and the second phase sample is a sub sample of the first phase sample

$$B[\hat{Y}_{RD}] = \left[N^2 \frac{(n'-n)}{n'n}\right] Y\left[C_{xx} - C_{xy}\right]$$

and

$$MSE[\hat{Y}_{RD}] = N^2 Y^2 \left[\frac{(N-n)}{Nn}\right]\left[C_{yy} + C_{xx} - 2C_{xy}\right] + \left[N^2 \frac{(N-n')}{Nn'}\right]\left[C_{xy} - C_{xx}\right]$$

The above theorems can be proved by applying the expressions given in (6.36) - (6.43) along with Theorems 6.8 - 6.11.

6.9 Use Of Multi-Auxiliary Information

There are many situations in which in addition to the study variable, information on several related auxiliary variables will be available. In such situations, the ratio estimator can be extended in several ways. In this section, one straight forward extension due to Olkin (1958) is considered.

Let $\hat{X}_i$ be unbiased for $X_i, i = 1, 2, ..., k$ the population total of the ith auxiliary variable and $\hat{Y}$ be unbiased for Y, the population total of the study variable. Olkin (1958) suggested a composite estimator of the form

$$\hat{Y}_{Rk} = \sum_{i=1}^{k} W_i \left[\frac{\hat{Y}}{\hat{X}_i}\right] X_i \tag{6.44}$$

where $W_1, W_2, ..., W_k$ are predetermined constants satisfying $\sum_{i=1}^{k} W_i = 1$.

Note that if k=2, the above estimator reduces to

$$\hat{Y}_{R2} = W_1 \frac{\hat{Y}}{\hat{X}_1} X_1 + W_2 \frac{\hat{Y}}{\hat{X}_2} X_2 \tag{6.45}$$

where $W_1 + W_2 = 1$.

The following theorem gives the approximate bias and mean square error of the estimator $\hat{Y}_{R2}$. In order to make the expressions compact, the following notations are used.

$$V_0 = \frac{V(\hat{Y})}{Y^2}, V_1 = \frac{V(\hat{X}_1)}{X_1^{\,2}}, V_2 = \frac{V(\hat{X}_2)}{X_2^{\,2}}, C_{01} = \frac{\text{cov}(\hat{Y}, \hat{X}_1)}{YX_1},$$

$$C_{02} = \frac{\text{cov}(\hat{Y}, \hat{X}_2)}{YX_2}, C_{12} = \frac{\text{cov}(X_1, \hat{X}_2)}{X_1 X_2}.$$

Theorem 6.14 The approximate bias and mean square error of $\hat{Y}_{R2}$ are

$$B(\hat{Y}_{R2}) = Y\{V_2 - C_{02} + W_1(C_{02} - C_{01} - V_2)\}$$

and

$$MSE(\hat{Y}_{R2}) = Y^2 \left\{ \begin{array}{l} V_0 + V_2 - 2C_{02} + W_1^2(V_2 + V_1 - 2C_{12}) - \\ \qquad 2W_1(C_{01} + V_2 - C_{02} - C_{12}) \end{array} \right\}$$

Proof Let $e_0 = \dfrac{\hat{Y} - Y}{Y}$, $e_1 = \dfrac{\hat{X}_1 - X_1}{X_1}$ and $e_2 = \dfrac{\hat{X}_2 - X_2}{X_2}$

The estimator $\hat{Y}_{R2}$ can be written as

$$\hat{Y}_{R2} = W_1 Y(1+e_0)(1+e_1)^{-1} + W_2 Y(1+e_0)(1+e_2)^{-1}$$

$$= Y\begin{Bmatrix} W_1(1+e_0)(1-e_1+e_1^2-...)+ \\ W_2(1+e_0)(1-e_2+e_2^2-...) \end{Bmatrix}$$

$$= Y\begin{Bmatrix} W_1(1-e_1+e_1^2-.e_0-e_0e_1+...)+ \\ (1-W_1)(1-e_2+e_2^2+e_0-e_0e_2+...) \end{Bmatrix}$$

$$= Y\begin{Bmatrix} (1-e_2+e_2^2+e_0-e_0e_2...)+W_1(1-e_1+e_1^2-.e_0-e_0e_1... \\ -1+e_2-e_2^2+e_0+e_0e_2...) \end{Bmatrix}$$

$$\text{Therefore } \hat{Y}_{R2} - Y = Y\begin{Bmatrix} (e_0-e_2+e_2^2-e_0e_2...)+ \\ W_1(e_2-e_1+e_1^2-e_0e_1-e_2^2+e_0e_2...) \end{Bmatrix} \tag{6.46}$$

Taking expectations on both the sides after ignoring terms of degree greater than two, we get the approximate bias

$$B(\hat{Y}_{R2}) = Y\{V_2 - C_{02} + W_1(C_{02} - C_{01} - V_2)\} \tag{6.47}$$

Squaring both the sides of (6.46) and taking expectation, on ignoring terms of degree greater than two, we get the approximate mean square error

$$MSE(\hat{Y}_{R2}) = Y^2\begin{Bmatrix} V_0+V_2-2C_{02}+W_1^2(V_2+V_1-2C_{12})- \\ 2W_1(C_{01}+V_2-C_{02}-C_{12}) \end{Bmatrix} \tag{6.48}$$

Hence the proof. ■

Remark Note that the mean square error given in (6.48) attains minimum if

$$W_1 = \frac{C_{01}+V_2-C_{02}-C_{12}}{V_2+V_1-2C_{12}} \tag{6.49}$$

The minimum mean square error of the estimator $\hat{Y}_{R2}$ obtained by substituting (6.49) in (6.48) is

$$Y^2\left\{V_0+V_2-2C_{02}-\frac{(C_{01}+V_2-C_{02}-C_{12})^2}{V_2+V_1-2C_{12}}\right\} \tag{6.50}$$

It is pertinent to note that the denominator of the expression is nothing but the variance of the difference $e_1 - e_2$. Therefore the minimum mean square error given in (6.50) is always less than or equal to $Y^2\{V_0+V_2-2C_{02}\}$ which is nothing but the approximate mean square error of the ratio estimator based on the auxiliary variable x_2. Therefore we infer by using the additional auxiliary variable the efficiency of the ratio estimator can be increased. It is to be noted that the optimum value of W_1 given in (6.49) requires the knowledge of some parametric values which in general will not be known in advance. The usual practice is using their estimated values.

6.10 Ratio Estimation in Stratified Sampling

When the sample is selected in the form of a stratified sample, the ratio estimator can be constructed in two different ways.

Let $\hat{Y}_h$ and $\hat{X}_h, h = 1, 2, \ldots, L$ be unbiased for the population totals Y_h and X_h, hth stratum totals of the study and auxiliary variables respectively. Using these estimates, the population total can be estimated by using any one of the following estimates :

$$\hat{Y}_{RS} = \sum_{h=1}^{L} \frac{\hat{Y}_L}{\hat{X}_L} X_h \tag{6.51}$$

$$\hat{Y}_{RC} = \frac{\sum_{h=1}^{L} \hat{Y}_h}{\sum_{h=1}^{L} \hat{X}_h} X \tag{6.52}$$

The estimates $\hat{Y}_{RS}$ and $\hat{Y}_{RC}$ are known as separate ratio estimator and combined ratio estimator respectively. The separate estimator can be used to estimate the population total only when the true stratum total X_h of the auxiliary variable is known for all strata.

Theorem 6.15 The approximate bias and mean square error of the separate ratio estimator are $B[\hat{Y}_{RS}] = \sum_{h=1}^{L} Y_h \left\{ \left[\frac{V(\hat{X}_h)}{X_h^2} \right] - \left[\frac{\text{cov}(\hat{X}_h, \hat{Y}_h)}{X_h Y_h} \right] \right\}$

and $MSE[\hat{Y}_{RS}] = \sum_{h=1}^{L} Y_h^2 \left\{ \left[\frac{V(\hat{Y}_h)}{Y_h^2} \right] + \left[\frac{V(\hat{X}_h)}{X_h^2} \right] - 2 \left[\frac{\text{cov}(\hat{X}_h, \hat{Y}_h)}{X_h Y_h} \right] \right\}$

Proof Bias of the estimator under consideration is

$$B[\hat{Y}_{RS}] = E[\hat{Y}_{RS}] - Y$$

$$= \sum_{h=1}^{L} E\left\{ \left[\frac{\hat{Y}_h}{\hat{X}_h} X_h \right] - Y_h \right\}$$

$$= \sum_{h=1}^{L} Y_h \left\{ \left[\frac{V(\hat{X}_h)}{X_h^2} \right] - \left[\frac{\text{cov}(\hat{X}_h, \hat{Y}_h)}{X_h Y_h} \right] \right\} \quad \text{(using Theorem 6.1)}$$

Hence the proof. ■

The mean square error of the separate ratio estimator is

$$MSE[\hat{Y}_{RS}] = E[\hat{Y}_{RS} - Y]^2$$

$$=\sum_{h=1}^{L}E\left\{\left[\frac{\hat{Y}_h}{\hat{X}_h}X_h\right]-Y_h\right\}^2$$

$$=\sum_{h=1}^{L}Y_h^2\left\{\left[\frac{V(\hat{Y}_h)}{Y_h^2}\right]+\left[\frac{V(\hat{X}_h)}{X_h^2}\right]-2\left[\frac{\text{cov}(\hat{X}_h,\hat{Y}_h)}{X_hY_h}\right]\right\} \tag{6.54}$$

The above mean square error is only an approximate expression and it is obtained by applying Theorem 6.1 under the assumptions stated in the same theorem.

The combined ratio estimator is constructed by using $\sum_{h=1}^{L}\hat{X}_h$ and $\sum_{h=1}^{L}\hat{Y}_h$ as estimates for the population totals X and Y respectively. Therefore the approximate bias and mean square error are

$$B[\hat{Y}_{RC}]=Y\frac{V(\sum_{h=1}^{L}\hat{X}_h)}{X^2}-Y\frac{\text{cov}\left(\sum_{h=1}^{L}\hat{X}_h,\sum_{h=1}^{L}\hat{Y}_h\right)}{XY}$$

$$=Y\frac{\sum_{h=1}^{L}V(\hat{X}_h)}{X^2}-Y\frac{\left(\sum_{h=1}^{L}\text{cov}(\hat{X}_h,\hat{Y}_h)\right)}{XY} \tag{6.55}$$

and $$MSE[\hat{Y}_{RC}]=Y^2\left\{\frac{V(\sum_{h=1}^{L}\hat{Y}_h)}{Y^2}+\frac{V(\sum_{h=1}^{L}\hat{X}_h)}{X^2}-2\frac{\text{cov}\left(\sum_{h=1}^{L}\hat{X}_h,\sum_{h=1}^{L}\hat{Y}_h\right)}{XY}\right\}$$

$$=Y^2\left\{\frac{\sum_{h=1}^{L}V(\hat{Y}_h)}{Y^2}+\frac{\sum_{h=1}^{L}V(\hat{X}_h)}{X^2}-2\frac{\sum_{h=1}^{L}\text{cov}(\hat{X}_h,\hat{Y}_h)}{XY}\right\} \tag{6.56}$$

respectively.

The expressions given in (6.53)-(6.56) are applicable for any sampling design. In particular, if simple random sampling is used in all the L strata then they reduce to (*i*) $B[\hat{Y}_{RS}]=\sum_{h=1}^{L}Y_h\frac{N_h^2(N_h-n_h)}{N_hn_h}\{C_{xxh}-C_{xyh}\}$ and

$$(ii)\ MSE[\hat{Y}_{RS}]=\sum_{h=1}^{L}Y_h^2\frac{N_h^2(N_h-n_h)}{N_hn_h}\{C_{xxh}+C_{yyh}-2C_{xyh}\}$$

6.11 Problems and Solutions

Problem 6.1 Consider the estimator $\hat{Y}_\alpha = \hat{Y}\left[\frac{X}{\hat{X}}\right]^\alpha$ which reduces to the ratio estimator when α = 1 and the conventional expansion estimator $\hat{Y}$ if α = 0. Derive the approximate bias and mean square error of the above estimator and also the minimum mean square error with respect to α (Shrivastava, 1967).

Solution Using the notations introduced in Section 6.2, the estimator $\hat{Y}_\alpha = \hat{Y}\left[\frac{X}{\hat{X}}\right]^\alpha$ can be written as

$$\begin{aligned}\hat{Y}_\alpha &= Y(1+e_0)(1+e_1)^{-\alpha}\\ &= Y(1+e_0)\left\{1-\alpha e_1+\frac{\alpha(\alpha+1)}{2}e_1^2-\ldots\right\}\\ &= Y\left\{1-\alpha e_1+\frac{\alpha(\alpha+1)}{2}e_1^2+e_0-\alpha e_0 e_1+\ldots\ldots\right\}\end{aligned}$$

Therefore
$$\hat{Y}_\alpha - Y = Y\left\{e_0-\alpha e_1+\frac{\alpha(\alpha+1)}{2}e_1^2+e_0-\alpha e_0 e_1+\ldots\right\} \tag{6.57}$$

Taking expectation on both the sides after ignoring terms of degree greater than two, we get the approximate bias as

$$\begin{aligned}B &= Y\left\{\frac{\alpha(\alpha+1)}{2}E(e_1^2)-\alpha E(e_0e_1)\right\}\\ &= Y\left\{\frac{\alpha(\alpha+1)}{2}\frac{V(\hat{X})}{X^2}-\alpha\frac{\operatorname{cov}(\hat{X},\hat{Y})}{XY}\right\}\end{aligned} \tag{6.58}$$

Squaring and taking expectations on both the sides of (6.57) after ignoring terms of degree greater than two, we get the approximate mean square error as

$$\begin{aligned}M &= [E(e_0^2)+\alpha^2E(e_1^2)-2\alpha E(e_0e_1)]\\ &= Y^2\left\{\frac{V(\hat{Y})}{Y^2}+\alpha^2\frac{V(\hat{X})}{X^2}-2\alpha\frac{\operatorname{cov}(\hat{Y},\hat{X})}{XY}\right\}\end{aligned} \tag{6.59}$$

By employing the usual calculus methods, we note that the above mean square error is minimum if

$$\alpha = \frac{X}{Y}\frac{\operatorname{cov}(\hat{X},\hat{Y})}{XY} \tag{6.60}$$

Substituting (6.60) in (6.59), we get the minimum mean square error as $V(\hat{Y})(1-\rho^2)$ where ρ is the correlation coefficient between $\hat{Y}$ and $\hat{X}$. ■

***Problem* 6.2** Consider the estimator $\hat{Y}_{\alpha} = \alpha\hat{Y} + (1-\alpha)\hat{Y}\dfrac{\hat{X}}{X}$ which reduces to the product and conventional expansion estimators when $\alpha = 1$ and 0 respectively.

Solution The estimator $\hat{Y}_{\alpha}$ can be written as

$$\hat{Y}_{\alpha} = \alpha\hat{Y} + (1-\alpha)\hat{Y}\frac{\hat{X}}{X}$$

$$= \alpha y + (1+e_0) + (1-\alpha)Y(1+e_0)(1+e_1)$$

$$= Y + (1+e_0) + [\alpha + (1-\alpha)(1+e_1)]$$

$$= Y[1 + e_1 - \alpha e_1 + e_0 + e_0 e_1 - \alpha e_0 e_1]$$

Therefore $\hat{Y}_{\alpha} - Y = Y[e_0 + (1+\alpha)(e_1 + e_0 e_1)]$

Taking expectations on both sides, we get the bias of the estimator

$$B = Y(1-\alpha)\left[\frac{\operatorname{cov}(\hat{X},\hat{Y})}{XY}\right]$$

Squaring both the sides of (6.61) and taking expectations after ignoring terms of degree greater than two, we get the approximate mean square error as

$$M = Y^2\left\{\frac{V(\hat{Y})}{Y^2} + (1-\alpha)^2\frac{V(\hat{X})}{X^2} + 2(1-\alpha)\frac{\operatorname{cov}(\hat{X},\hat{Y})}{XY}\right\}$$

It can be seen that the above mean square error is minimum if

$$\alpha = 1 + \left[\frac{X}{Y}\right]\left[\frac{\operatorname{cov}(\hat{X},\hat{Y})}{XY}\right]$$

and the minimum mean square error is $V(\hat{Y})(1-\rho^2)$ where ρ is the correlation coefficient between $\hat{Y}$ and $\hat{X}$. ■

***Problem* 6.3** Derive the approximate bias and mean square error of the estimator $\alpha\hat{Y}\dfrac{X}{\hat{X}}$ and also find the minimum mean square error.

Solution The given estimator can be written as

$$\hat{Y}_{\alpha} = \alpha\hat{Y}\frac{X}{\hat{X}}$$

$$= \alpha Y(1+e_0)(1+e_1)^{-1}$$

$$= \alpha Y(1+e_0)(1-e_1+e_1^2-\ldots)$$

$$= \alpha Y(1-e_1+e_1^2+e_0-e_0e_1+..)$$

Therefore $\hat{Y}_{\alpha} - Y = (\alpha-1)Y - \alpha Y(e_0 - e_1 + e_1^2 + e_0 - e_0 e_1 + \ldots)$

Taking expectations on both sides of the above expression we get the bias of the estimator as

$$(\alpha-1)Y-\alpha Y\left\{\frac{V(\hat{X})}{X^2}-\frac{\text{cov}(\hat{X},\hat{Y})}{XY}\right\}$$

Proceeding in the usual way we get the mean square error of the estimator as

$$M=(\alpha-1)^2+\alpha^2Y^2\left\{\frac{V(\hat{Y})}{Y^2}+\frac{V(\hat{X})}{X^2}-2\frac{\text{cov}(\hat{Y},\hat{X})}{XY}\right\}$$

$$-2\alpha(\alpha-1)Y^2\left\{\frac{V(\hat{X})}{X^2}-\frac{\text{cov}(\hat{X},\hat{Y})}{XY}\right\}$$

Minimising the above mean square error with respect to α and substituting the optimum value in the mean square error expression we get the minimum mean square error. ■

Problem 6.4 Let $\bar{x}_1$ and $\bar{x}$ be samples means in two phase sampling when samples are drawn independently and simple random sampling is used in both the phases of sampling. Show that the estimator

$$\bar{y}^*=\overline{r}\bar{x}_1+\frac{n(N-1)}{N(n-1)}(\bar{y}-\overline{r}\bar{x})$$

is unbiased for the population mean where $\bar{y}$ is the sample mean of the study variable based on the second phase sample.

Solution Since the samples are drawn without replacement and independently of each other in the two phases of sampling we have

$$E[\bar{y}^*]=E[\bar{r}]E[\bar{x}_1]+\frac{n(N-1)}{N(n-1)}E(\bar{y}-\bar{r}\,\bar{x})$$

Note that $E[\bar{r}\,\bar{x}]=\text{cov}(\bar{r},\bar{x})+E[\bar{r}]E[\bar{x}]$

$$=\frac{N-n}{Nn}\frac{1}{N-1}\sum_{i=1}^{N}(R_i-\overline{R})(X_i-\overline{X})+\overline{R}\,\overline{X}$$

$$=\frac{N-n}{n(N-1)}\overline{Y}+\frac{(n-1)N}{n(N-1)}\overline{R}\,\overline{X}\text{ where }\overline{R}=\frac{1}{N}\sum_{i=1}^{N}R_i,R_i=\frac{Y_i}{X_i}$$

Therefore we have $E(\bar{y}^*)=\dfrac{n(N-1)}{N(n-1)}\dfrac{N(n-1)}{n(N-1)}(\overline{Y}-\overline{R}\,\overline{X})+\overline{R}\,\overline{X}$

$$=\overline{Y}$$

Hence the solution. ■

Problem 6.5 Derive an unbiased ratio-type estimator based on $\bar{x}_w$, the mean of say w distinct units in two phase sampling when independent samples are drawn in the two phases of sampling using simple random sampling.

Solution Let $\bar{r}\,\bar{x}_w$ be an estimator of the population mean $\overline{Y}$ where $\bar{r}$ is as defined in the last problem. Note that $E[\bar{r}\,\bar{x}_w]=E\{E(\bar{r}\,\bar{x}_w\mid w)\}$

Since each units in a particular subset s_w (containing w distinct units) is given equal chance for being included in the sample, we get $E[\bar{r} \mid w] = \bar{r}_w$.

Therefore

$$E[\bar{r}\,\bar{x}_w] = E[\bar{r}_w \bar{x}_w]$$

$$= \text{cov}(\bar{r}_w, \bar{x}_w) + E[\bar{r}_w]E[\bar{x}_w]$$

$$= E\left\{\frac{N-w}{Nw}\frac{1}{N-1}\sum_{i=1}^{N}(R_i - \bar{R})(X_i - \bar{X})\right\} + \bar{R}\,\bar{X} \text{ (refer the next problem)}$$

where the expectation in the right hand side of the above expression is with respect to the distribution of w.

$$E[\bar{r}\,\bar{x}_w] = E\left\{\frac{N-w}{Nw}\frac{1}{N-1}\right\}\bar{Y} - E\left\{\frac{N(w-1)}{w(N-1)}\right\}\bar{R}\,\bar{X}$$

Hence the bias of $\bar{r}\,\bar{x}_w$ is $B(\bar{r}\,\bar{x}_w \mid w) = E(\bar{r}\,\bar{x}_w \mid w) - \bar{Y}$

$$= -E\left\{\frac{N(w-1)}{w(N-1)}\right\}(\bar{Y} - \bar{R}\,\bar{X})$$

Further we know that $E[\bar{y} - \bar{r}\,\bar{x}] = \left\{\frac{N(n-1)}{n(N-1)}\right\}(\bar{Y} - \bar{R}\,\bar{X})$

Therefore $B(\bar{r}\,\bar{x}_w) = -E\left\{\frac{N(w-1)}{w(N-1)}\right\}\left\{\frac{n(N-1)}{N(n-1)}\right\}(\bar{y} - \bar{r}\,\bar{x})$

$$= -E\left\{\frac{n(w-1)}{w(n-1)}\right\}(\bar{y} - \bar{r}\,\bar{x})$$

Hence an unbiased estimator of the population mean is

$$\bar{y}_w = \overline{rx}_w + \left\{\frac{n(w-1)}{w(n-1)}\right\}(\bar{y} - \bar{r}\,\bar{x}) \quad \blacksquare$$

Problem 6.6 Under the notations used in problem 6.5 derive (a) $E(\bar{x}_w)$, (b) $V(\bar{x}_w)$ and (c) $\text{cov}(\bar{x}_w, \bar{y})$

Solution (a) $E(\bar{x}_w) = EE(\bar{x}_w \mid w)$

$$= E(\bar{X})$$

$$= \bar{X}$$

(b) $V(\bar{x}_w) = EV(\bar{x}_w \mid w) + VE(\bar{x}_w \mid w)$

$$= E\left\{\frac{1}{w} - \frac{1}{N}\right\}S_x^2 + V(\bar{X})$$

$$= \left\{E\left(\frac{1}{w}\right) - \frac{1}{N}\right\}S_x^2$$

(c) We can write $E(\bar{y} \mid s_w) = \bar{y}_w$

Therefore $\text{cov}(\bar{x}_w, \bar{y}) = E(\bar{x}_w \bar{y}) - E(\bar{x}_w)E(\bar{y})$

$$= EE(\bar{x}_w \bar{y} \mid s_w) - E(\bar{x}_w)E(\bar{y} \mid s_w)$$

$$= \text{cov}(\bar{x}_w, \bar{y}_w)$$

$$= \left\{ E\left(\frac{1}{w}\right) - \frac{1}{N} \right\} S_{xy}$$

Hence the solution. ■

Exercises

6.1 Derive under simple random sampling the approximate bias and mean square error of the estimator $\hat{Y}_{RS} = \hat{Y} \frac{S_x^2}{s_x^2}$.

6.2 Derive the approximate bias and minimum mean square error of the estimator $\hat{Y} = \frac{\hat{Y}}{X}[aX + (1-a)\hat{X}]$ and compare the minimum mean square error with the mean square error of the Linear regression estimator.

6.3 Let $\bar{x}^*$ be the mean of distinct units when samples are drawn independently in two phase sampling. Derive the approximate bias and mean square error of the estimator $\hat{\bar{Y}}^* = \frac{\bar{Y}}{\bar{\bar{X}}} \bar{x}^*$ assuming simple random sampling is used in both the phases of sampling.

6.4 Derive the minimum square error of the estimator $\hat{\bar{Y}}^* = \hat{\bar{Y}} + b(\bar{x}^* - \bar{x})$ under the notations explained in 6.3.

6.5 Show that for a sample of n units selecting using *srswor*,

$$B[\hat{R}_n \bar{X}] = \frac{n}{N} \frac{N-1}{n-1} E[\bar{x}(\hat{R}_n - \hat{R}_1)]$$

where $\hat{R}_1 = \frac{\bar{y}}{\bar{x}}$ and $\hat{R}_n = \frac{1}{n} \sum_{i=1}^{n} \frac{y_i}{x_i}$, y_i and x_i being y and x values of the ith drawn unit.

6.6 Let C_x be the coefficient of variation of x. Derive the condition under which the estimator $\hat{Y} \frac{\bar{X} + C_x}{\bar{X} - C_x}$ is more efficient than the usual ratio estimator, assuming simple random sampling is used.

Chapter 7

Regression Estimation

7.1 Introduction

Like ratio estimation, regression estimation is another method of estimation of a finite population total using the knowledge of an auxiliary variable x which is closely related to the study variable y. The regression estimator is developed below.

We know that, when the variables x and y are linearly related, the least squares estimates of the slope and intercept are respectively $\hat{b} = \frac{s_{xy}}{s_x^2}$ and $\hat{a} = \hat{\overline{Y}} - \hat{b}\hat{\overline{X}}$.

The Y can be expressed as

$$Y = \sum_{i \in s} Y_i + \sum_{i \in \overline{s}} Y_i \text{ where } \overline{s} = S - s. \tag{7.1}$$

Once the sample is observed, the first term in the right hand side becomes fully known. Using the least squares estimates $\hat{b} = \frac{s_{xy}}{s_x^2}$ and $\hat{a} = \hat{\overline{Y}} - \hat{b}\hat{\overline{X}}$ each unobserved y value can be estimated by

$$\begin{aligned} \hat{Y}_i &= \hat{a} + \hat{b}X_i, i \in \overline{s} \\ &= \hat{\overline{Y}} - \hat{b}\hat{\overline{X}} + \hat{b}X_i \end{aligned}$$

Summing both the sides over $\overline{s} = S - s$, we get

$$\begin{aligned} \sum_{i \in \overline{s}} \hat{Y}_i &= (N-n)(\hat{\overline{Y}} - \hat{b}\hat{\overline{X}}) + \hat{b}\sum_{i \in \overline{s}} X_i \\ &= (N-n)(\hat{\overline{Y}} - \hat{b}\hat{\overline{X}}) + \hat{b}[X - \sum_{i \in \overline{s}} X_i] \\ &= (N-n)(\hat{\overline{Y}} - \hat{b}\hat{\overline{X}}) + \hat{b}[N\overline{X} - n\hat{\overline{X}}] \end{aligned}$$

Substituting these estimated values for the unobserved y values in (7.1), we get an estimator for the population total Y as

$$\hat{Y} = \sum_{i \in s} Y_i + (N-n)(\hat{\overline{Y}} - \hat{b}\hat{\overline{X}}) + \hat{b}[N\overline{X} - n\hat{\overline{X}}]$$

$$= n\hat{\bar{Y}} + (N-n)(\hat{\bar{Y}} - \hat{b}\hat{\bar{X}}) + \hat{b}[N\overline{X} - n\hat{\bar{X}}]$$
$$= N\hat{\bar{Y}} + N\hat{b}[N\overline{X} - n\hat{\bar{X}}] \tag{7.2}$$

The above estimator is known as the Linear Regression estimator of the population total Y. It should be noted that the above estimator is not unbiased for the population total under simple random sampling. The following theorem gives the approximate mean square error of the Linear regression estimator.

Theorem 7.1 The approximate mean square error of the regression estimator under simple random sampling is $N^2 \dfrac{(N-n)}{Nn} S_y^2(1-\rho^2)$

Proof Define $e_0 = \dfrac{\hat{\bar{Y}} - \overline{Y}}{\overline{Y}}, e_1 = \dfrac{\hat{\bar{X}} - \overline{X}}{\overline{X}}, e_2 = \dfrac{s_{xy} - S_{xy}}{S_{xy}}, e_3 = \dfrac{s_x^2 - S_x^3}{S_x^2}$

It is to be noted that $E(e_i) = 0, i = 0, 1, 2, 3$.

The regression estimator can be expressed as

$$\hat{Y}_{LR} = N\hat{\bar{Y}} + N\hat{b}[N\overline{X} - n\hat{\bar{X}}]$$
$$= Y(1+e_0) + \frac{S_{xy}(1+e_2)}{S_x^2(1+e_3)}[X - X(1+e_1)]$$
$$= Y(1+e_0) - \frac{XS_{xy}}{S_x^2}(1+e_2)(1+e_3)^{-1}e_1$$
$$= Y(1+e_0) - XB(1+e_2)(1+e_3)^{-1}e_1 \text{ where } B = \frac{S_{xy}}{S_x^2}$$

Assuming $|e_i| < 1, i = 0, 1, 2, 3$, the above expression can be modified as

$$\hat{Y}_{LR} - Y = Ye_0 - XBe_1(1+e_2)(1 - e_3 + e_3^2 - e_3^4 + \ldots)$$
$$= Ye_0 - XB(e_1 - e_1e_3 + e_1e_2) \text{ (ignoring terms of degree greater than two)}$$

Squaring both the sides and taking expectations we get

$$E(\hat{Y}_{LR} - Y)^2 = Y^2E(e_0^2) + X^2B^2E(e_1^2) - 2XYBE(e_0e_1)$$
$$= \frac{N^2(N-n)}{Nn}[S_y^2 + B^2S_x^2 - 2BS_{xy}]$$
$$= \frac{N^2(N-n)}{Nn}[S_y^2 + \frac{S_{xy}^2}{S_x^4}S_x^2 - 2\frac{S_{xy}}{S_x^2}S_{xy}]$$
$$= \frac{N^2(N-n)}{Nn}S_y^2\left[1 - \frac{S_{xy}^2}{S_x^2S_y^2}\right]$$
$$= N^2\frac{(N-n)}{Nn}S_y^2(1-\rho^2)$$

Hence the proof. ■

Theorem 7.2 Under simple random sampling, $V[\hat{Y}_{srs}] > MSE[\hat{Y}_{LR}]$ and $MSE[\hat{Y}_R] > MSE[\hat{Y}_{LR}]$.

Proof Since $-1 < \rho < 1$, we have $(1-\rho^2) < 1$

Therefore $N^2 \frac{(N-n)}{Nn} S_y^2 (1-\rho^2) < N^2 \frac{(N-n)}{Nn} S_y^2$

Hence $V[\hat{Y}_{srs}] > MSE[\hat{Y}_{LR}]$.

Consider the difference

$$MSE[\hat{Y}_R] - MSE[\hat{Y}_{LR}] = N^2 \frac{(N-n)}{Nn} \{S_y^2 + R^2 S_x^2 - 2RS_{xy} - S_y^2 + S_y^2 \rho^2\}$$

$$= N^2 \frac{(N-n)}{Nn} \{S_y^2 \rho^2 + R^2 S_x^2 - 2RS_{xy}\}$$

$$= N^2 \frac{(N-n)}{Nn} \left[\frac{S_y^2 S_{xy}^2}{S_x^2 S_y^2} + R^2 S_x^2 - 2RS_{xy} \right]$$

$$= N^2 \frac{(N-n)}{Nn} \frac{(S_{xy} - RS_x)^2}{S_x^2}$$

Since the right hand side of the above expression is always non-negative, the result follows. ■

7.2 Difference Estimation

The ratio estimator which is obtained by multiplying the conventional estimator $\hat{Y}$ by the factor $\frac{X}{\hat{X}}$ is an alternative to the estimator $\hat{Y}$. Here we shall examine the possibility of improving upon $\hat{Y}$ by considering the estimator obtained by adding $\hat{Y}$ with constant times the difference $X - \hat{X}$ whose expected value is zero. That is, as an estimator of Y, we take

$$\hat{Y}_{DR} = \hat{Y} + \lambda(X - \hat{X}) \tag{7.3}$$

where λ is a predetermined value. Since the above estimator depends on the difference $X - \hat{X}$ rather than the ratio $\frac{X}{\hat{X}}$, it is termed as "Difference estimator". The difference estimator is unbiased for the population total Y and its variance is

$$V(\hat{Y}_{DR}) = E[\hat{Y}_{DR} - Y]^2$$

$$= E[(\hat{Y} - Y) + \lambda(X - \hat{X})]^2$$

$$= E(\hat{Y} - Y)^2 + \lambda^2 E(X - \hat{X})^2 - 2\lambda E(\hat{X} - X)(\hat{Y} - Y)$$

$$= V(\hat{Y}) + \lambda^2 V(\hat{X}) - 2\lambda \operatorname{cov}(\hat{X}, \hat{Y}) \tag{7.4}$$

The above expression for variance is applicable for any sampling design yielding unbiased estimator for Y and X. It can be seen that the above variance is minimum if

$$\lambda = \frac{\operatorname{cov}(\hat{X}, \hat{Y})}{XY} \tag{7.5}$$

and the resulting minimum variance is $V(\hat{Y})[1-\rho^2(\hat{X},\hat{Y})]$ where $\rho(\hat{X},\hat{Y})$ is the coefficient of correlation between $\hat{X}$ and $\hat{Y}$. It is interesting to note that, when simple random sampling is used, the optimum value of λ is $\frac{S_{xy}}{S_x^2}$ and the minimum variance happens to be $N^2\frac{(N-n)}{Nn}S_y^2(1-\rho^2)$ which is nothing but the approximate mean square error of the linear regression estimator. It is pertinent to note that the optimum value of λ depends on S_{xy} which in general will not be known. Normally in such situations, survey practitioners use unbiased estimators for unknown quantities. The value derived from the optimal choice happens to be the least squares estimate. Therefore the estimator $\hat{Y}_{DR}$ reduces to the linear regression estimator. It is also to be noted that the difference estimator reduces to the ratio estimator when $\lambda = \frac{\hat{Y}}{\hat{X}}$.

7.3 Double Sampling in Difference Estimation

As in the case of ratio estimation, here also one can employ double sampling method to estimate the population total Y whenever the population total X of the auxiliary variable is not known. The difference estimator for the population total under double sampling is defined as

$$\hat{Y}_{DD} = \hat{Y} + \lambda(\hat{X}_d - \hat{X}) \tag{7.6}$$

where $\hat{X}_D$ is an unbiased estimator of the population total X based on the first phase sample. Evidently the difference estimator is unbiased for the population total in both the cases of double sampling.

Note that $V(\hat{Y}_{DD}) = E[\hat{Y}_{DD} - Y]^2$

$$\begin{aligned} &= E[(\hat{Y}-Y)+\lambda(\hat{X}_d-\hat{X})]^2 \\ &= E[(\hat{Y}-Y)+\lambda[(\hat{X}_d-X)-(\hat{X}-X)]]^2 \\ &= V(\hat{Y})+\lambda^2[V(\hat{X})+V(\hat{X}_d)-2\operatorname{cov}(\hat{X},\hat{X}_d)] \\ &\quad -2\lambda[\operatorname{cov}(\hat{Y},\hat{X}_d)-\operatorname{cov}(\hat{X},\hat{Y})] \end{aligned} \tag{7.7}$$

When the samples are drawn independently, the above variance reduces to

$$V[\hat{Y}_{DD}] = V(\hat{Y}) + \lambda^2[V(\hat{X})+V(\hat{X}_d)] - 2\lambda\operatorname{cov}(\hat{Y},\hat{X}_d)$$

The following theorem gives the variance of the difference estimator in double sampling when the samples are drawn independently in two phases of sampling using simple random sampling.

Theorem 7.3 When the samples are drawn independently in the two phases of sampling using simple random sampling the variance of the difference estimator is

$$V(\hat{Y}_{DD}) = N^2[S_y^2 + \lambda^2(f + f')S_x^2 - 2\lambda f S_{xy}]$$

where $f = \dfrac{N-n}{Nn}$ and $f' = \dfrac{N-n'}{Nn'}$. Here n' and n are the sample sizes corresponding to the first and second phases of sampling. Further the minimum variance of the difference estimator in this case is

$$N^2 f S_y^2\left[1 - \frac{f}{f+f'}\rho^2\right]$$

where ρ is the correlation coefficient between x and y.

Proof Using the results stated in Section 6.8 in the variance expression available in (7.7) we get

$$V(\hat{Y}_{DD}) = \frac{N^2(N-n)}{Nn}S_y^2 + \lambda^2\frac{N^2(N-n')}{Nn'}S_x^2 + \frac{N^2(N-n)}{Nn}S_x^2$$
$$-2\lambda\frac{N^2(N-n)}{Nn}S_{xy}$$
$$= N^2[S_y^2 + \lambda^2(f + f')S_x^2 - 2\lambda f S_{xy}] \quad (7.8)$$

Differentiating the above variance expression partially with respect to λ and equating the derivative to zero, we get $\lambda = \dfrac{f}{f+f'}\dfrac{S_{xy}}{S_x^2}$. Substituting this value in (7.8) and simplifying the resulting expression we get the minimum variance

$$N^2 f S_y^2\left[1 - \frac{f}{f+f'}\rho^2\right]$$

It is to be noted that the second order derivative is always positive. Hence the proof. ■

Theorem 7.4 When the second phase sample is a subsample of the first phase sample and simple random sampling is used in both the phases of sampling the variance of the difference estimator is

$$V(\hat{Y}_{DD}) = N^2[f S_y^2 + \lambda^2(f - f')S_x^2 + 2\lambda(f' - f)S_{xy}].$$

The minimum variance of the difference estimator in this case is

$$N^2 S_y^2[f'\rho^2 + f(1-\rho^2)]$$

where f and f' are as defined in Theorem 7.3

Proof of this theorem is left as an exercise.

7.4 Multivariate Difference Estimation

When information about more than one auxiliary variable is known, the difference estimator defined in Section 7.3 can be extended in a straight forward manner.

Let $\hat{Y}$, $\hat{X}_1$ and $\hat{X}_2$ be unbiased estimators for the population totals Y, X_1 and X_2 of the study variable y, the auxiliary variables x_1 and x_2 respectively. The difference estimator of the population total Y is defined as

$$\hat{Y}_{D2} = \hat{Y} + B_1(X_1 - \hat{X}_1) + B_2(X_2 - \hat{X}_2) \tag{7.9}$$

where the constants B_1 and B_2 are predetermined.

The estimator $\hat{Y}_{D2}$ is unbiased for the population total and its variance is

$$\begin{aligned} V(\hat{Y}_{D2}) &= E[(\hat{Y} - Y) + B_1(X_1 - \hat{X}_1) + B_2(X_2 - \hat{X}_2)]^2 \\ &= V(\hat{Y}) + B_1^2 V(\hat{X}_1) + B_2^2 V(\hat{X}_2) - 2B_1 \operatorname{cov}(\hat{Y}, \hat{X}_1) \\ &\qquad - 2B_2 \operatorname{cov}(\hat{Y}, \hat{X}_2) + 2B_1 B_2 \operatorname{cov}(\hat{X}_1, \hat{X}_2) \end{aligned}$$

Denote by

$$V_0 = V(\hat{Y}), V_1 = V(\hat{X}_1), V_2 = V(\hat{X}_2)$$

$$C_{01} = \operatorname{cov}(\hat{Y}, \hat{X}_1), C_{02} = \operatorname{cov}(\hat{Y}, \hat{X}_2), C_{12} = \operatorname{cov}(\hat{X}_1, \hat{X}_2)$$

Differentiating the variance expression partially with respect to B_1 and B_2 and equating the derivatives to zero, we get the following equations

$$V_1 B_1 + C_{12} B_2 = C_{01} \tag{7.10}$$

$$C_{12} B_1 + V_2 B_2 = C_{02} \tag{7.11}$$

Solving these two equations, we obtain

$$B_1 = \frac{C_{01} V_2 - C_{12} C_{02}}{V_1 V_2 - C_{12}^2} \tag{7.12}$$

$$B_2 = \frac{C_{02} V_1 - C_{12} C_{01}}{V_1 V_2 - C_{12}^2} \tag{7.13}$$

Substituting these values in the variance expression, we get after simplification

$$V(\hat{Y})[1 - R^2_{y.x_1,x_2}] \tag{7.14}$$

where $R_{y.x_1,x_2}$ is the multiple correlation between $\hat{Y}$ and $\hat{X}_1$, $\hat{X}_2$. Since the multiple correlation between $\hat{Y}$ and $\hat{X}_1, \hat{X}_2$ is always greater than the correlation between $\hat{Y}$ and $\hat{X}_1$ and that of between $\hat{Y}$ and $\hat{X}_2$, we infer that the use of additional auxiliary information will always increase the efficiency of the estimator. However, it should be noted that the values of B_1 and B_2 given in (7.12) and (7.13) depend on C_{01} and C_{02} which in general will not be known. The following theorem proves that whenever b_1 and b_2 are used in place of B_1 and B_2 given in (7.12) and (7.13), the resulting estimator will have mean square error that is approximately equal to the minimum variance given in (7.14), where

$b_1 = \frac{c_{01}V_2 - C_{12}c_{02}}{V_1V_2 - C_{12}^2}$ and $b_2 = \frac{c_{02}V_1 - C_{12}c_{01}}{V_1V_2 - C_{12}^2}$ where c_{01} and c_{02} are unbiased estimators for C_{01} and C_{02} respectively.

Theorem 7.5 The approximate mean square error of the estimator

$$\hat{Y}_{D2}^* = \hat{Y} + b_1(X_1 - \hat{X}_1) + b_2(X_2 - \hat{X}_2)$$

is same as that of the difference estimator defined in (7.9), where b_1 and b_2 are as defined in (7.15) and (7.16) respectively.

Proof Let

$$e_0 = \frac{\hat{Y} - Y}{Y}, e_1 = \frac{\hat{X}_1 - X_1}{X_1}, e_2 = \frac{\hat{X}_2 - X_2}{X_2},$$

$$e' = \frac{c_{01} - C_{01}}{C_{01}}, e'' = \frac{c_{02} - C_{02}}{C_{02}}$$

It can be seen that

$$b_1 = \frac{C_{01}(1+e')V_2 - C_{12}C_{02}(1+e'')}{V_1V_2 - C_{12}^2}$$

$$= B_1 + \frac{C_{01}V_2e' - C_{12}C_{02}e''}{V_1V_2 - C_{12}^2} \quad (7.17)$$

Similarly it be seen that

$$b_2 = B_2 + \frac{C_{02}V_1e'' - C_{12}C_{01}e'}{V_1V_2 - C_{12}^2} \quad (7.18)$$

Using (7.17) and (7.18), the estimator $\hat{Y}_{D2}^*$ can be written as

$$\hat{Y}_{D2}^* = \hat{Y} + \left[B_1 + \frac{C_{01}V_2e' - C_{12}C_{02}e''}{V_1V_2 - C_{12}^2}\right](X_1 - \hat{X}_1) + \left[B_2 + \frac{C_{02}V_1e'' - C_{12}C_{01}e'}{V_1V_2 - C_{12}^2}\right](X_2 - \hat{X}_2) \quad (7.19)$$

Replacing $\hat{Y}$, $\hat{X}_1$ and $\hat{X}_2$ by $Y(1+e_0)$, $X_1(1+e_1)$ and $X_2(1+e_2)$ in (7.19) we obtain

$$\hat{Y}_{D2}^* - Y = Ye_0 + \left[B_1 + \frac{C_{01}V_2e' - C_{12}C_{02}e''}{V_1V_2 - C_{12}^2}\right](-X_1e_1) + \left[B_2 + \frac{C_{02}V_1e'' - C_{12}C_{01}e'}{V_1V_2 - C_{12}^2}\right](-X_2e_2)$$

Squaring both the sides and ignoring terms of degree greater than two, we obtain

$$E[\hat{Y}_{D2}^* - Y]^2 = Y^2E(e_0^2) + B_1^2X_1^2E(e_1^2) + B_2^2X_2^2E(e_2^2) - 2B_1YX_1E(e_0e_1) - 2B_2YX_2E(e_0e_2) + 2B_1B_2X_1X_2E(e_1e_2)$$

$$= V(\hat{Y}) + B_1^2 V(\hat{X}_1) + B_2^2 V(\hat{X}_2) -$$

$$2B_1 \operatorname{cov}(\hat{Y}, \hat{X}_1) - 2B_2 \operatorname{cov}(\hat{Y}, \hat{X}_2) + 2B_1 B_2 \operatorname{cov}(\hat{X}_1, \hat{X}_2)$$

Substituting the values given in (7.12) and (7,13) in the above expression, we get the approximate mean square error of $\hat{Y}_{D2}^*$ as $V(\hat{Y})[1 - R^2_{y.x_1,x_2}]$. Hence the proof. ■

Thus in the last few sections of this chapter, we constructed the linear regression estimator for the population total by using the fact that the variables x and y are linearly related and extended this to cover the case of having more than one auxiliary variable. In the following section, the problem of identifying the optimal sampling-estimating strategy with the help of super population models is considered.

7.5 Inference under Super-population Models

In the super-population approach, the population values are assumed to be the realised values of N independent random variables. In this section, we shall assume that the population values are the realised values of N independent random variables $Y_1, Y_2, \ldots, Y_N$ where Y_i has mean bx_i and variance $\sigma^2 v(x_i)$. The function $v(.)$ is known, $v(x) \geq 0$ for $x \geq 0$. The constants b and σ^2 are unknown. Let ξ denote the joint probability law.

Implied estimator Consider any estimator $\hat{T}$ for the population total T which can be uniquely expressed in the form $\hat{T} = \sum_s Y_i + \hat{b} \sum_{\bar{s}} X_i$ where $\hat{b}$ does not depend on the unobserved y-values. Here $\hat{b}$ is referred to as the implied estimator of the parameter b.

For example, under simple random sampling, the ratio estimator can be expressed as $\hat{T}_R = \frac{\sum_s Y_i}{\sum_s X_i} \sum_{i=1}^N X_i = \sum_s Y_i + \frac{\sum_s Y_i}{\sum_s X_i} \sum_{\bar{s}} X_i$. Therefore, the implied estimator of b corresponding to the ratio estimator is $\frac{\sum_s Y_i}{\sum_s X_i}$. It is to be noted that the implied estimator does not depend on unobserved y's. In the following theorem we shall prove that of two statistics $\hat{T}$ and $\hat{T}^0$, the one whose implied estimator for b is better is, the better estimator for T.

Theorem 7.6 For any sampling design P, if estimators $\hat{T}$ and $\hat{T}^0$ have implied estimators $\hat{b}$ and $\hat{b}^0$ for b which satisfy

$$E_\xi[\hat{b} - b]^2 \leq E_\xi[\hat{b}^0 - b]^2 \tag{7.21}$$

for each s such that $P(s)>0$, then

$$MSE(P:\hat{T}) \le MSE(P:\hat{T}^0) \tag{7.22}$$

If for some subset, say s_0 of S with $P(s_0)>0$, the inequality in (7.21) is strict, the strict inequality holds in (7.22).

Proof Under the super-population model described earlier,

$$MSE(P:\hat{T}) = E_\xi[\sum_s P(s)(\hat{T}-T)^2] \tag{7.23}$$

and $$MSE(P:\hat{T}^0) = E_\xi[\sum_s P(s)(\hat{T}^0-T)^2] \tag{7.24}$$

Therefore $MSE(P:\hat{T}) \le MSE(P:\hat{T}^0)$

$$E_\xi[\sum_s P(s)(\hat{T}-T)^2] \le E_\xi[\sum_s P(s)(\hat{T}^0-T)^2] \tag{7.25}$$

Consider
$$\begin{aligned}\hat{T}-T &= \sum_s Y_i + \hat{b}\sum_{\bar{s}} X_i - \sum_s Y_i - \sum_{\bar{s}} Y_i \\ &= \hat{b}\sum_{\bar{s}} X_i - \sum_{\bar{s}} Y_i \\ &= (\hat{b}-b)\sum_{\bar{s}} X_i - [\sum_{\bar{s}}(Y_i - bX_i)]\end{aligned}$$

Squaring both the sides and taking expectations, we obtain

$$\begin{aligned}E_\xi[\hat{T}-T]^2 = {} & [\sum_{\bar{s}} X_i]^2 E_\xi(\hat{b}-b)^2 + E_\xi[\sum_{\bar{s}}(Y_i - bX_i)]^2 \\ & - 2[\sum_{\bar{s}} X_i]E_\xi(\hat{b}-b)^2 \sum_{\bar{s}}(Y_i - bX_i)]\end{aligned} \tag{7.26}$$

Since $\hat{b}$ is independent of unobserved y's and $E_\xi(\hat{b}-b)=0$,

$$E_\xi[\hat{b}-b]\sum_{\bar{s}}(Y_i - bX_i)] = E_\xi[\hat{b}-b]\sum_{\bar{s}} E_\xi(Y_i - bX_i)] = 0 \tag{7.27}$$

Further
$$\begin{aligned}E_\xi[\sum_{\bar{s}}(Y_i - bX_i)]^2 &= \sum_{\bar{s}} E_\xi(Y_i - bX_i)^2 \text{ (since } y \text{ s'are independent)} \\ &= \sigma^2 \sum_{\bar{s}} v(X_i)\end{aligned} \tag{7.28}$$

Substituting (7.27) and (7.28) in (7.26) we get

$$E_\xi[\hat{T}-T]^2 = [\sum_{\bar{s}} X_i]^2 E_\xi(\hat{b}-b)^2 + \sigma^2 \sum_{\bar{s}} v(X_i) \tag{7.29}$$

Proceeding in the same way, we obtain

$$E_\xi[\hat{T}^0-T]^2 = [\sum_{\bar{s}} X_i]^2 E_\xi(\hat{b}^0-b)^2 + \sigma^2 \sum_{\bar{s}} v(X_i) \tag{7.30}$$

If $E_\xi[\hat{b}-b]^2 \le E_\xi[\hat{b}^0-b]^2$ for each s such that $P(s)>0$, then

$$[\sum_{\bar{s}} X_i]^2 E_\xi (\hat{b}-b)^2 + \sigma^2 \sum_{\bar{s}} v(X_i)$$

$$\leq [\sum_{\bar{s}} X_i]^2 E_\xi (\hat{b}^0 - b)^2 + \sigma^2 \sum_{\bar{s}} v(X_i)$$

$\Rightarrow \; E_\xi [\hat{T}-T]^2] \leq E_\xi [\hat{T}^0 - T]^2$ for each s such that $P(s)>0$.

Hence the proof follows from (7.25) ■

Thus we conclude that the estimator which has the better implied estimator is better. In order to make the study more deeper under the super-population approach, we give the following definition.

Model Unbiasedness An estimator $\hat{T}$ is said to be model unbiased or ξ unbiased if for each s, $\hat{T} - \sum_s y_i$ is an unbiased predictor of the unobserved sum $\sum_{\bar{s}} y_i$. That is, if $E_\xi [\hat{T} - \sum_s y_i] = E_\xi [\sum_{\bar{s}} y_i]$ for every s (7.31)

It may be seen that the above definition is equivalent to $E_\xi [\hat{T} - T] = 0$.

***Theorem** 7.7* An estimator is model unbiased if and only if the implied estimator is unbiased for b

Proof We know that $\hat{T} - T = (\hat{b} - b) \sum_{\bar{s}} X_i - [\sum_{\bar{s}} (Y_i - bX_i)]$ (refer (7.25))

Taking expectations on both the sides we get

$$E_\xi (\hat{T} - T) = \sum_{\bar{s}} X_i E_\xi (\hat{b} - b)$$

From this we infer that $E_\xi (\hat{T} - T) = 0 \Leftrightarrow E_\xi (\hat{b} - b) = 0$.

Hence the proof. ■

Note A design unbiased estimator is not necessarily model unbiased and in the same way a model unbiased estimator is not necessarily design unbiased. The following theorem gives the Best Linear Model unbiased estimator for the population total.

Theorem 7.8 For any sampling design P, let $\hat{T}$ be a linear estimator satisfying $E_\xi (\hat{T} - T) = 0$ for ever s such that $P(s)>0$, then $MSE(P:\hat{T}^*) \leq MSE(P:\hat{T})$

where $\hat{T}^* = \sum_s Y_i + \hat{b}^* \sum_{\bar{s}} X_i$, $\hat{b}^* = \dfrac{\sum_s \frac{X_i Y_i}{v(X_i)}}{\sum_s \frac{X_i^2}{v(X_i)}}$

Proof Since we consider only estimators that can be expressed in the form

$$\hat{T} = \sum_s Y_i + \hat{b}\sum_{\bar{s}} X_i$$

it is easily seen that $\hat{T}$ is a linear function of the observed y's if and only if $\hat{b}$ is a linear function of the observed y's. Therefore by Gauss-Markov Theorem and Theorem 7.6, the proof follows. ■

The above theorem helps us to derive the Best Linear Unbiased estimators (with respect to the model) for different choices of $v(.)$.

Case (1) : $v(x) = 1$

$$\hat{b}^* = \frac{\sum_s \frac{X_i Y_i}{1}}{\sum_s \frac{X_i^2}{1}} = \frac{\sum_s X_i Y_i}{\sum_s X_i^2} \tag{7.32}$$

Case (2): $v(x) = x$

$$\hat{b}^* = \frac{\sum_s \frac{X_i Y_i}{X_i}}{\sum_s \frac{X_i^2}{X_i}} = \frac{\sum_s Y_i}{\sum_s X_i} \tag{7.33}$$

Case (3): $v(x) = x^2$

$$\hat{b}^* = \frac{\sum_s \frac{X_i Y_i}{X_i^2}}{\sum_s \frac{X_i^2}{X_i^2}} = \frac{1}{n}\sum_s \frac{Y_i}{X_i} \tag{7.34}$$

Thus under the three cases the Best Linear Unbiased Estimators are

$$\hat{T}_1 = \sum_s Y_i + \frac{\sum_s X_i Y_i}{\sum_s X_i^2}\sum_{\bar{s}} X_i, \hat{T}_2 = \sum_s Y_i + \frac{\sum_s Y_i}{\sum_s X_i}\sum_{\bar{s}} X_i \text{ and } \hat{T}_3 = \sum_s Y_i +$$

$\frac{1}{n}\sum_s \frac{Y_i}{X_i}\sum_{\bar{s}} X_i$ respectively. It is interesting to note that the estimator $\hat{T}_2$ is nothing but the ratio estimator. This proves the ratio estimator is the Best Linear Unbiased Estimator for the population total when $v(x) = x$.

Now we shall state and prove a lemma which will be used later to prove an interesting property regarding the estimator $\hat{T}_3 = \sum_s Y_i + \frac{1}{n}\sum_s \frac{Y_i}{X_i}\sum_{\bar{s}} X_i$.

Lemma 7.1 If $0 \le b_1 \le b_2 \le ... \le b_m$ and if $c_1 \le c_2 \le ... \le c_m$ satisfies $c_1 + c_2 + ... + c_m \ge 0$ then $b_1c_1 + b_2c_2 + ... + b_mc_m \ge 0$

Proof Let k denote the greatest integer i for which $c_i \leq 0$. Then

$$\begin{aligned} b_1c_1 + b_2c_2 + \ldots + b_mc_m &= \sum_{i=1}^{k} b_ic_i + \sum_{i=k+1}^{m} b_ic_i \\ &\geq b_k \sum_{i=1}^{k} c_i + b_{k+1} \sum_{i=k+1}^{m} c_i \\ &\geq b_k \sum_{i=1}^{m} c_i + (b_{k+1} - b_k) \sum_{i=k+1}^{m} c_i \\ &\geq 0 \end{aligned}$$

Hence the proof. ■

The following theorem proves $\hat{T}_3$ is better than $\hat{T}_{pps} = \frac{X}{n}\sum_s \frac{Y_i}{X_i}$ for any fixed size sampling design under a wide class of variance functions.

Theorem 7.9 If $Max\{nX_1, nX_2, \ldots, nX_N\} \leq \sum_{i=1}^{N} X_i$ and $v(x)$ is non-increasing in x then for any sampling plan P for which $P(s)>0$ only if $n(s)=n$ then $MSE(P:\hat{T}_{pps}) \leq MSE(P:\hat{T}_3)$.

Proof In order to prove the given result, it is enough to show that

$$E_\xi[\hat{b}_{pps} - b]^2 \leq E_\xi[\hat{b}_3 - b]^2$$

where $\hat{b}_{pps}$ is the implied estimator of b corresponding to the estimator $\hat{T}_{pps}$. To identify the implied estimator corresponding to the estimator $\hat{T}_{pps}$, it can be written as

$$\hat{T}_{pps} = \sum_s Y_i + \frac{\frac{X}{n}\sum_s \frac{Y_i}{X_i} - \sum_s Y_i}{\sum_{\bar{s}} X_i} \sum_{\bar{s}} X_i$$

Therefore the implied estimator corresponding to the estimator $\hat{T}_{pps}$ is

$$\begin{aligned} \hat{b}_{pps} &= \frac{\frac{X}{n}\sum_s \frac{Y_i}{X_i} - \sum_s Y_i}{\sum_{\bar{s}} X_i} \\ &= \sum_s \frac{Y_i a_i}{nX_i} \end{aligned}$$

where $a_i = \dfrac{X - nX_i}{\sum_{\bar{s}} X_i}$

Hence $E_\xi[\hat{b}_{pps} - b]^2 = E_\xi\left[\sum_s \dfrac{Y_i a_i}{nX_i} - b\right]^2$

$$= E_\xi\left[\sum_s \frac{a_i}{nX_i}[Y_i - E_\xi(Y_i)]\right]^2$$

$$= \frac{\sigma^2}{n^2}\sum_s \frac{a_i^2 v(X_i)}{X_i^2} \tag{7.35}$$

Further note that $\hat{b}_3 - b = \dfrac{1}{n}\sum_s \dfrac{Y_i}{X_i} - b$

$$= \frac{1}{n}\sum_s \frac{1}{X_i}[Y_i - E_\xi(Y_i)]$$

Squaring both the sides and taking expectations, we get

$$E_\xi[\hat{b}_3 - b]^2 = \frac{1}{n^2}\sum_s \frac{1}{X_i^2} E_\xi[Y_i - E_\xi(Y_i)]^2$$

$$= \frac{\sigma^2}{n^2}\sum_s \frac{v(X_i)}{X_i^2} \tag{7.36}$$

From (7.35) and (7.36) we get

$$E_\xi[\hat{b}_{pps} - b]^2 - E_\xi[\hat{b}_3 - b]^2 = \frac{\sigma^2}{n^2}\sum_s \frac{(a_i^2 - 1)v(X_i)}{X_i^2} \tag{7.37}$$

From the definition of a_i, we get $a_i^2 \geq a_j^2 \Rightarrow X_i \leq X_j$. Therefore whenever $a_i^2 - 1 \geq a_j^2 - 1$, we note that $\dfrac{v(X_i)}{X_i^2} \geq \dfrac{v(X_j)}{X_j^2}$. Further under the given conditions $\sum_s [a_i^2 - 1] \geq 0$. Therefore by the Lemma 7.1 the proof follows. ■

Thus we have proved the estimator $\hat{T}_3$ is better than $\hat{T}_{pps}$. It is to be noted that the estimator $\hat{T}_{pps}$ is not model unbiased even though it is linear which makes the above theorem meaningful. Thus we have identified the optimal estimators under the given super-population model for different choices of the variance function appearing in the super-population model. Now we shall come to the problem of identifying the ideal sampling plan with respect to the given super-population model.

If we fix the sample size, then the problem of identifying an optimal sampling design becomes straight forward. Let S_n denote the set of all subsets of size n of the population S and P_n be the collection of all sampling designs P for which $P(s)>0$ only when s is in S_n. Since for P in P_n,

$$MSE(P:\hat{T}) = E_\xi[\sum_s P(s)(\hat{T}-T)^2] \tag{7.37}$$

clearly the optimal sampling plan is one which selects with certainty a subset s which minimises $E_\xi(\hat{T}-T)^2$. Some insight can be gained when the quantity to be minimised is expressed in the form

$$E_\xi[\hat{T}-T]^2 = [\sum_{\bar{s}} X_i]^2 E_\xi(\hat{b}-b)^2 + \sigma^2 \sum_{\bar{s}} v(X_i) \tag{7.38}$$

From (7.38) we understand that the sampler has two objectives namely (1) to choose a sample which will afford a good estimate of the expected value of the total of the non-sampled values. That is, to choose a sample s so that $[\sum_{\bar{s}} X_i]^2 E_\xi(\hat{b}-b)^2$ is small and (2) to observe those units whose values of y have greatest variance so that only the sum of the least variable values must be predicted. That is to choose s so that $\sigma^2 \sum_{\bar{s}} v(X_i)$ is small.

For a wide class of variance functions, if the optimal estimator defined in Theorem 7.8 is to be used, then determination of the optimal sample is quite simple. As it is shown in the following theorem, the best sample to observe consists of those units having the largest x values. Let s^* be any sample of n units for which $Max_{S_n} \sum_s X_i = \sum_{s^*} X_i$ and p^* be a sampling plan which entails selecting s^* with certainty. That is, $P^*(s=s^*)=1$.

Theorem 7.10 If $\frac{v(X)}{X^2}$ is non-increasing then $MSE(P^*:\hat{T}^*) \le MSE(P:\hat{T}^*)$ for any sampling plan P in P_n where

$$\hat{T}^* = \sum_s Y_i + \hat{b}^* \sum_{\bar{s}} X_i, \text{ and } \hat{b}^* = \frac{\sum_s \frac{X_i Y_i}{v(X_i)}}{\sum_s \frac{X_i^2}{v(X_i)}}$$

Proof We know that

$$MSE(P:\hat{T}) = E_\xi[\sum_s P(s)(\hat{T}-T)^2]$$

and $$E_\xi[\hat{T}-T]^2 = [\sum_{\bar{s}} X_i]^2 E_\xi(\hat{b}-b)^2 + \sigma^2 \sum_{\bar{s}} v(X_i)$$

Note that $(\hat{b}-b)=\dfrac{\sum_s \dfrac{X_iY_i}{v(X_i)}}{\sum_s \dfrac{X_i^2}{v(X_i)}}-b$

$$=\frac{\sum_s \dfrac{X_i[Y-bX_i]}{v(X_i)}}{\sum_s \dfrac{X_i^2}{v(X_i)}}$$

Since y's are independent, squaring both the sides and taking expectations we get

$$E_\xi(\hat{b}-b)^2=\frac{\sigma^2\sum_s\left[\dfrac{X_i}{v(X_i)}\right]^2 v(X_i)}{\left[\sum_s \dfrac{X_i^2}{v(X_i)}\right]^2}$$

$$=\frac{\sigma^2}{\left[\sum_s \dfrac{X_i^2}{v(X_i)}\right]}$$

Therefore $MSE(P:\hat{T})=\sigma^2\sum_{S_n}P(s)\left[\sum_{\bar{s}}v(X_i)+\dfrac{\left(\sum_{\bar{s}}v(X_i)\right)^2}{\sum_s \dfrac{X_i^2}{v(X_i)}}\right]$

Clearly the expression in side the square brackets is minimum for $s=s^*$. Since the sampling plan P^* selects the sample s^* with certainty we have $MSE(P^*:\hat{T}^*)\le MSE(P:\hat{T}^*)$. Hence the proof.

The following theorem proves the sampling plan P* is optimal for use with the estimator $\hat{T}_3$ and $\hat{T}_{pps}$ under rather weak conditions.

Theorem 7.11 For any P in P_n and any $v(.)$ for which both $v(X)$ and $\dfrac{X^2}{v(X)}$ are non-decreasing

(i) $MSE(P^*:\hat{T}_{ppa})\le MSE(P:\hat{T}_{pps})$ if $Max\{nX_1,nX_2,\ldots,nX_N\}\le\sum_{i=1}^{N}X_i$

(ii) $MSE(P^*:\hat{T}_3)\le MSE(P:\hat{T}_3)$

Proof From (7.35) we have

$$E_\xi[\hat{b}_{pps}-b]^2=\frac{\sigma^2}{n^2}\sum_s\frac{a_i^2 v(X_i)}{X_i^2} \quad \text{where} \quad a_i=\frac{X-nX_i}{\sum_{\bar{s}}X_i}$$

Therefore $$E_\xi[\hat{b}_{pps}-b]^2=\frac{1}{\left(\sum_{\bar{s}}X_i\right)^2}\frac{\sigma^2}{n^2}\sum_s\frac{v(X_i)}{X_i^2}[X-nX_i]^2$$

$$E_\xi[\hat{T}_{pps}-T]^2=\sigma^2\sum_{\bar{s}}v(X_i)+\frac{\sigma^2}{n^2}\sum_s\frac{v(X_i)}{X_i^2}[X-nX_i]^2$$

Hence

$$MSE(P:\hat{T}_{pps})=\sum_{S_n}P(s)\{\sigma^2\sum_{\bar{s}}v(X_i)+\frac{\sigma^2}{n^2}\sum_s\frac{v(X_i)}{X_i^2}[X-nX_i]^2\}$$

It can be seen that the above expression will be minimum when $s=s^*$. Since P^* is the sampling plan which selects s^* with certainty

$$MSE(P^*:\hat{T}_{ppa})\leq MSE(P:\hat{T}_{pps})$$

This completes the first part.

We have seen in Theorem 7.9, $E_\xi[\hat{b}_3-b]^2=\frac{\sigma^2}{n^2}\sum_s\frac{v(X_i)}{X_i^2}$

Therefore $$E_\xi[\hat{T}_3-T]^2=\left(\sum_{\bar{s}}X_i\right)^2\frac{\sigma^2}{n^2}\sum_s\frac{v(X_i)}{X_i^2}+\sigma^2\sum_{\bar{s}}v(X_i)$$

Hence $$MSE(P:\hat{T}_3)=\sum_{S_n}P(s)\{\sigma^2\sum_{\bar{s}}v(X_i)+\left(\sum_{\bar{s}}X_i\right)^2\frac{\sigma^2}{n^2}\sum_s\frac{v(X_i)}{X_i^2}\}$$

Clearly under the assumptions stated in the theorem, the right hand side of the above expression is minimum if $s=s^*$. Since the sampling plan P^* yields the set s^* as sample with probability one,

$$MSE(P^*:\hat{T}_3)\leq MSE(P:\hat{T}_3)$$

Hence the proof. ■

The results discussed in this section are due to Royall (1971).

7.6 Problems and Solutions

Problem 7.1 Derive the Best Linear Unbiased Estimator for the population total under the super-population model

$$E_\xi[Y_i]=a+bi, i=1,\ 2,\ \ldots,\ N\ ,\ V_\xi[Y_i]=\sigma^2 v(i)\,;\mathrm{cov}_\xi[Y_i,Y_j]=0, i\neq j$$

where ξ is the joint probability law of $Y_1,Y_2,\ldots,Y_N$. Also find its mean square error when $a=0$.

Solution By Theorem 7.8, the Best Linear Unbiased Estimator for the population total is given by $\hat{T} = \sum_s Y_i + (N-n)\hat{a} + \hat{b}\sum_{\bar{s}} X_i$ where

$$\hat{a} = \frac{\sum_s \frac{Y_i}{v(i)} - \hat{b}\sum_s \frac{i}{v(i)}}{\sum_s \frac{i}{v(i)}} \quad \text{and} \quad \hat{b} = \frac{\sum_s \frac{iY_i}{v(i)}\sum_s \frac{i}{v(i)} - \sum_s \frac{Y_i}{v(i)}\sum_s \frac{i}{v(i)}}{\sum_s \frac{i^2}{v(i)}\sum_s \frac{1}{v(i)} - \left[\sum_s \frac{i}{v(i)}\right]^2}$$

When $a=0$, the estimator reduces to $\hat{T}^* = \sum_s Y_i + \hat{b}^* \sum_{\bar{s}} X_i$, $\hat{b}^* = \dfrac{\sum_s \frac{iY_i}{v(i)}}{\sum_s \frac{i^2}{v(i)}}$

Consider $\quad (\hat{b}^* - b) = \dfrac{\sum_s \frac{iY_i}{i}}{\sum_s \frac{i^2}{v(i)}} - b$

$$= \frac{\sum_s \frac{i[Y_i - bi]}{v(i)}}{\sum_s \frac{i^2}{v(i)}}$$

Since y's are independent, squaring both the sides and taking expectations we get

$$E_\xi(\hat{b}^* - b)^2 = \frac{\sigma^2 \sum_s \left[\frac{i}{v(i)}\right]^2 v(i)}{\left[\sum_s \frac{i^2}{v(i)}\right]^2}$$

$$= \frac{\sigma^2}{\left[\sum_s \frac{i^2}{v(i)}\right]} \qquad (*)$$

One can write $\hat{T}^* - T = \sum_s Y_i + \hat{b}^* \sum_{\bar{s}} i - \sum_s Y_i - \sum_{\bar{s}} Y_i$

$$= (\hat{b}^* - b)\sum_{\bar{s}} i - \sum_{\bar{s}} [Y_i - E_\xi(Y_i)]$$

Squaring and taking expectation with respect to the model, we get

$$E_\xi(\hat{T}^* - T)^2 = E_\xi(\hat{b}^* - b)^2 [\sum_{\bar{s}} i]^2 + \sum_{\bar{s}} \sigma^2 v(i) \qquad (**)$$

Using (*) in (**) we get

$$E_\xi(\hat{T}^* - T)^2 = \frac{\sigma^2}{\left[\sum_s \frac{i^2}{v(i)}\right]} [\sum_{\bar{s}} i]^2 + \sum_{\bar{s}} \sigma^2 v(i)$$

Hence by definition of mean square error, we get

$$MSE(P:\hat{T}^*) = \sum_{S_n} P(s) \frac{\sigma^2}{\left[\sum_s \frac{i^2}{v(i)}\right]} [\sum_{\bar{s}} i]^2 + \sum_{\bar{s}} \sigma^2 v(i)$$

$$= \sigma^2 \left[\sum_{S_n} \left\{ \frac{\left(\sum_{\bar{s}} i\right)^2}{\sum_s \frac{i^2}{v(i)}} + \sum_{\bar{s}} v(i) \right\} P(s) \right]$$

Exercises

7.1 Show that the mean square error of $\hat{T}_0$ derived in the above problem is minimum in P_n under the sampling design $P_0(s)$, where

$$P_0(s) = \begin{cases} 1 \text{ if } s = s^* \\ 0 \text{ otherwise} \end{cases}$$

where s* is the set containing the units with labels $N-n+1$, $N-n+2$, ...,*N*, provided the function $v(i)$ is non-decreasing in i and $\frac{v(i)}{i^2}$ is non-increasing in *i*. Here P_n is the class of sampling designs yielding samples of size *n*.

7.2 Extend the difference estimator considered in this chapter to *p*-auxiliary variables case and show that the resulting minimum mean square error is $\frac{N^2(N-n)}{Nn} S_y^2(1 - R_{0.123...p}^2)$, where $R_{0.123...p}$ is the multiple correlation coefficient, assuming simple random sampling is used.

7.3 Derive the Best Linear Unbiased Estimator for the population total under the super-population model,

$$E_\xi[Y_i] = a + bi + c_i^2, i = 1, 2, \ldots, N, V_\xi[Y_i] = \sigma^2;$$

$$\mathrm{cov}_\xi[Y_i, Y_j] = 0, i \neq j,$$

where ξ is the joint probability law of $Y_1, Y_2, \ldots, Y_N$. Also derive the mean square error of the estimator.

Chapter 8

Multistage sampling

8.1 Introduction

So far we have seen a number of sampling methods wherein a sample of units to be investigated are taken directly from the given population. While this is convenient in small scale surveys, it is not so in large scale surveys. The main reason being that no usable list describing the population units to be considered generally exists to select the sample. Even if such a list is available, it would be economically viable to base the enquiry on a simple random sample or systematic sample, because this would force the interviewer to visit almost each and every part of the population. Therefore it becomes necessary to select clusters of units rather than units directly from the given population. One way of selecting the sample would be to secure a list of clusters, take a probability sample of clusters and observe every unit in the sample. This is called **single-stage cluster sampling**. For example to estimate the total yield of wheat in a district during a given season, instead of treating individual fields as sampling units, one can treat clusters of neighbouring fields as sampling units and instead of selecting a sample of fields one can select clusters of fields. Sometimes instead of observing every field within each cluster, one can select samples of fields within each cluster. This is called **two-stage sampling** since now the sample is selected in two stages- first the cluster of fields (called first stage or primary stage units) and then the fields within the clusters. This is also called **Subsampling.** Generally, subsampling is done independently from all the selected primary units.

8.2 Estimation Under Cluster Sampling

Suppose the population is divided into N clusters where the ith cluster contains $N_i, (i = 1, 2, ..., N)$ units. Let $Y_{ij}\,(j = 1, 2, ..., N_i; i = 1, 2, ..., N)$ be the y-value of the jth unit in the ith cluster and $Y_i = \sum_{j=1}^{N_i} Y_{ij}$. That is, $Y_i\ (i = 1, 2, ..., N)$ stands for the total of all the units in the cluster i. Suppose a cluster sample of n clusters is drawn by using a sampling design with first order inclusion probabilities $\pi_i, (i = 1, 2, ..., N)$ and second order inclusion probabilities $\pi_{ij}, i \neq j$. An unbiased estimator for the population total Y of all the units in the population namely, $Y = \sum_{i=1}^{N}\sum_{j=1}^{N_i} Y_{ij}$ is given by

$$\hat{Y}_{cl} = \sum_{i\in s}\frac{Y_i}{\pi_i} \tag{8.1}$$

and its variance is

$$V(\hat{Y}_{cl}) = \sum_{i=1}^{N} Y_i^2\left[\frac{1-\pi_i}{\pi_i}\right] + 2\sum_{i=1}^{N}\sum_{\substack{j=1\\ i<j}}^{N} Y_iY_j\left[\frac{\pi_{ij}-\pi_i\pi_j}{\pi_i\pi_j}\right] \tag{8.2}$$

The expressions given in (8.1) and (8.2) can be used for estimating the population total and to get its variance under any sampling design for which the first and second order inclusion probabilities are known. In particular, when simple random sampling is used to get a sample of n clusters, then an unbiased estimator of the population total is given by

$$\hat{Y}_{cls} = \frac{N}{n}\sum_{i\in s} Y_i \tag{8.3}$$

and its variance is

$$V[\hat{Y}_{cls}] = \left[\frac{N^2(N-n)}{Nn}\right]S_y^2 \tag{8.4}$$

where $S_y^2 = \frac{1}{N-1}\sum_{i=1}^{N}[Y_i - \bar{Y}]^2$.

In the same manner, for all sampling designs appropriate unbiased estimators can be constructed and their variances can also be obtained.

It is interesting to note that the number of units in each cluster can be taken as a measure of size and cluster sampling can be performed with the help of probability proportional to size scheme. Let y_i be the total of the ith sampled cluster and p_i be the selection probability of the unit selected in the ith draw, $i = 1, 2, \ldots, n$ when a probability proportional to size sample of size n is drawn with replacement. Note that, when the number of units in each cluster is regarded as size, the selection probability of the rth unit in the population is given by $P_r = \frac{N_r}{N_0}, r = 1, 2, \ldots, N$, where $N_0 = \sum_{i=1}^{N} N_i$. In this case an unbiased estimator of the population total is given by

$$\hat{Y}_{clp} = \frac{1}{n}\sum_{i=1}^{n}\left[\frac{y_i}{p_i}\right] \tag{8.5}$$

and its variance is

$$V(\hat{Y}_{clp}) = \frac{1}{n}\sum_{i=1}^{N}\left\{\frac{Y_i}{P_i} - Y\right\}^2 P_i \tag{8.6}$$

Thus we have understood that no new principles are involved in constructing estimators when a probability sample of clusters is taken. A problem to be considered is the optimum size of cluster. No general solution is available for this problem. However, when clusters are of the same size and simple random sampling is used, partial answer is provided to this problem. The following theorem gives the variance of the estimator under simple random sampling in terms of intracluster correlation, when the clusters contain same number of units.

Theorem 8.1 When the clusters contain M units and a cluster sample of n clusters is drawn using simple random sampling, the variance of the estimator considered in (8.3) is

$$V[\hat{Y}_{cls}] = \left[\frac{N^2(N-n)}{Nn}\right]\left[\frac{NM-1}{N-1}\right] S_y^2[1+(M-1)\rho]$$

where ρ is the intracluster correlation given by

$$\rho = \frac{2\sum_{i=1}^{N}\sum_{j<k}^{M}[Y_{ij}-\overline{\overline{Y}}][Y_{ik}-\overline{\overline{Y}}]}{(M-1)(NM-1)S_y^2}, \quad \overline{\overline{Y}} = \frac{Y}{NM}$$

Proof We have seen in (8.4)

$$V[\hat{Y}_{cls}] = \left[\frac{N^2(N-n)}{Nn}\right]\frac{1}{N-1}\sum_{i=1}^{N}[Y_i-\overline{\overline{Y}}]^2 \tag{8.7}$$

Note that

$$\sum_{i=1}^{N}[Y_i-\overline{\overline{Y}}]^2 = \sum_{i=1}^{N}\left[\sum_{j=1}^{M}[Y_{ij}-\overline{\overline{Y}}]\right]^2$$

$$= \sum_{i=1}^{N}\left[\sum_{j=1}^{M}[Y_{ij}-\overline{\overline{Y}}]^2\right] + 2\sum_{i=1}^{N}\sum_{j<k}^{M}[Y_{ij}-\overline{\overline{Y}}][Y_{ik}-\overline{\overline{Y}}]$$

$$= (NM-1)S_y^2 + (M-1)(NM-1)\rho S_y^2$$

$$= (NM-1)S_y^2[1+(M-1)\rho] \tag{8.8}$$

Substituting (8.8) in (8.7) we get the required result. ■

From the above theorem we infer that the variance expression obtained depends on the number of clusters in the sample, the variance S_y^2, the size of the cluster M and the intracluster correlation coefficient.

Remark Since $\frac{(NM-1)}{N-1} \approx M$, the variance expression given in Theorem 8.1 can be written as

$$V[\hat{Y}_{cls}] = \left[\frac{N^2(N-n)}{Nn}\right] MS_y^2[1+(M-1)\rho] \tag{8.9}$$

Under the conditions stated in Theorem 8.1, if instead of sampling of clusters, a simple random sample of *nM* elements be taken directly from the population which contains *NM* elements,

$$V[\hat{Y}_{srs}] = \left[\frac{(NM)^2(NM-nM)}{M^2Nn}\right] S_y^2$$

$$= \left[\frac{N^2(N-n)}{Nn}\right] MS_y^2 \tag{8.10}$$

Comparing (8.9) with (8.10) we note that

$$V[\hat{Y}_{cls}] = V(\hat{Y}_{srs})[1+(M-1)\rho] \tag{8.11}$$

Since ρ is generally positive (because clusters are usually formed by putting together geographically contiguous elements), we infer from (8.11) cluster sampling will give a higher variance than sampling elements directly from the population. But it should be remembered that cluster sampling will be more economical when compared to simple random sampling. However, if ρ is negative, both the cost and the efficiency point to the use of cluster sampling.

8.3 Multistage Sampling

If the population contains very large number of units, we resort to sampling in several stages. For the first stage, we define a new population whose units are clusters of the original units. The clusters used for the first stage of sampling are called primary stage units *(psu)*. For example when the population is a collection of individuals living in a city, the psu may be taken as streets. Each psu selected in the first stage may be considered as a smaller population from which we select a certain number of smaller units, namely secondary stage units *(ssu)*. Unless stated otherwise, the second-stage sampling in each psu is carried out independently.

1. Two-stage sampling with simple random sampling in both the stages

As in the cluster of sampling, let Y_{ij} $(j = 1, 2, ..., N_i; i = 1, 2, ..., N)$ denote the *y*-value of the *j*th unit in the *i*th cluster (*psu*), ($\overline{Y}_{i.} = \frac{1}{N_i}\sum_{j=1}^{N_i} Y_{ij}$ mean per second-stage unit of the *i*th primary stage unit and $\overline{Y}_{..} = \frac{1}{N_i}\sum_{j=1}^{N} \overline{Y}_{i.}$ the population mean.

Assume that a sample of *n* primary stage units are selected using simple random sampling in the first stage and a subsample of size n_i is drawn from the *i*th primary stage sampled unit $i \in J$, J being the set of indices of the sampled primary stage units. The following theorem gives an unbiased estimator for the population total under two-stage sampling and also its variance.

Theorem 8.2 An unbiased estimator of the population total Y is given by

$\hat{Y}_{ms} = \frac{N}{n} \sum_{i \in J} N_i \bar{y}_{i.}$ $\bar{y}_{i.}$ being the mean of the units sampled from the ith sampled

psu and its variance is $V(\hat{Y}_{ms}) = \frac{N}{n} \sum_{i=1}^{N} N_i^2 \left[\frac{1}{n_i} - \frac{1}{N_i}\right] S_{wi}^2 + N^2 \left[\frac{1}{n} - \frac{1}{N}\right] S_b^2$

where $S_{wi}^2 = \frac{1}{M_i - 1} \sum_{j=1}^{N_i} [Y_{ij} - \bar{Y}_{i.}]^2$ and $S_b^2 = \frac{1}{N-1} \sum_{i=1}^{N} [\bar{Y}_{i.} - \bar{Y}_{..}]^2$

Proof $E(\hat{Y}_{ms}) = E_1 E_2(\hat{Y}_{ms})$, where E_1 and E_2 are the overall and conditional expectation with respect to subsampling respectively.

$$
\begin{aligned}
E(\hat{Y}_{ms}) &= E_1 E_2 \left[\frac{N}{n} \sum_{i \in J} N_i \bar{y}_{i.}\right] \\
&= E_1 \left[\frac{N}{n} \sum_{i \in J} N_i E_2(\bar{y}_{i.})\right] \\
&= E_1 \left[\frac{N}{n} \sum_{i \in J} N_i \bar{Y}_{i.}\right] \\
&= E_1 \left[\frac{N}{n} \sum_{i \in J} N_i Y_i\right] \\
&= Y
\end{aligned}
$$

The variance of $\hat{Y}_{ms}$ is given by

$$V(\hat{Y}_{ms}) = E_1 V_2(\hat{Y}_{ms}) + V_1 E_2(\hat{Y}_{ms}) \tag{8.12}$$

Note that $E_2(\hat{Y}_{ms}) = E_2 \left[\frac{N}{n} \sum_{i \in J} N_i \bar{y}_{i.}\right]$

$$
\begin{aligned}
&= \left[\frac{N}{n} \sum_{i \in J} N_i E_2(\bar{y}_{i.})\right] \\
&= \left[\frac{N}{n} \sum_{i \in J} N_i \bar{Y}_{i.}\right]
\end{aligned}
$$

Therefore $V_1E_2(\hat{Y}_{ms}) = N^2\left[\frac{1}{n}-\frac{1}{N}\right]S_b^2$ (8.13)

Further $V_2(\hat{Y}_{ms}) = \frac{N^2}{n^2}\sum_{i\in J} N_i^2\left[\frac{1}{n_i}-\frac{1}{N_i}\right]S_{wi}^2$

$$E_1V_2(\hat{Y}_{ms}) = \frac{N^2}{n^2}\frac{n}{N}\sum_{i=1}^{N} N_i^2\left[\frac{1}{n_i}-\frac{1}{N_i}\right]S_{wi}^2$$

$$= \frac{N}{n}\sum_{i=1}^{N} N_i^2\left[\frac{1}{n_i}-\frac{1}{N_i}\right]S_{wi}^2 \qquad (8.14)$$

Using (8.13) and (8.14) in (8.12), we get the required result. ■

The following theorem gives an unbiased estimator of the variance given in the above Theorem.

Theorem 8.3 An unbiased estimator of $V(\hat{Y}_{ms})$ is

$$v(\hat{Y}_{ms}) = \frac{N}{n}\sum_{i\in J} N_i^2\left[\frac{1}{n_i}-\frac{1}{N_i}\right]s_{wi}^2 + N^2\left[\frac{1}{n}-\frac{1}{N}\right]s_b^2$$

where $s_b^2 = \frac{1}{n-1}\sum_{i\in J}[N_i\bar{y}_{i.} - \frac{1}{n}\sum_{i\in J} N_i\bar{y}_{i.}]^2$ and $s_{wi}^2 = \frac{1}{n_i-1}\sum_{j=1}^{n_i}[y_{ij}-\bar{y}_{i.}]^2$,

y_{ij} being the value of the jth sampled unit from the ith sampled *psu*.

Proof We have

$$\sum_{i\in J}[N_i\bar{y}_{i.} - \frac{1}{n}\sum_{i\in J} N_i\bar{y}_{i.}]^2 = \sum_{i\in J} N_i^2\bar{y}_i^2 - \frac{1}{n}\left[\sum_{i\in J} N_i\bar{y}_i\right]^2 \qquad (8.15)$$

Consider $E\left[\sum_{i\in J} N_i^2\bar{y}_i^2\right] = E_1E_2\left[\sum_{i\in J} N_i^2\bar{y}_i^2\right]$

$$= E_1\left[\sum_{i\in J}\{N_i^2V_2(\bar{y}_i) + N_i^2[E_2(\bar{y}_{i.})]^2\}\right]$$

$$= E_1\left[\sum_{i\in J}\{N_i^2\left\{\frac{1}{n_i}-\frac{1}{N_i}\right\}S_{wi}^2 + N_i^2\bar{Y}_{i.}^2\}\right]$$

$$= \frac{n}{N}\left[\sum_{i=1}^{N} N_i^2\left\{\frac{1}{n_i}-\frac{1}{N_i}\right\}S_{wi}^2 + \frac{n}{N}\sum_{i=1}^{N} Y_i^2\right] \qquad (8.16)$$

Further

$$E\left[\sum_{i\in J} N_i\bar{y}_i\right]^2 = E_1E_2\left[\sum_{i\in J} N_i\bar{y}_i\right]^2$$

$$= E_1\left\{\left[E_2\sum_{i\in J} N_i\bar{y}_i\right]^2 + V_2\left[\sum_{i\in J} N_i\bar{y}_i\right]\right\}$$

$$= E_1\left\{\left[\sum_{i\in J} N_i\bar{Y}_i\right]^2 + \left[\sum_{i\in J} N_i^2\left(\frac{1}{n_i}-\frac{1}{N_i}\right)S_{wi}^2\right]\right\}$$

$$= E_1\left\{\left[\sum_{i\in J} Y_i\right]^2 + \frac{n}{N}\left[\sum_{i=1}^{N} N_i^2\left(\frac{1}{n_i}-\frac{1}{N_i}\right)S_{wi}^2\right]\right\}$$

$$= n^2\bar{Y}^2 + n^2\left[\frac{1}{n}-\frac{1}{N}\right]S_b^2 + \frac{n}{N}\left[\sum_{i=1}^{N} N_i^2\left(\frac{1}{n_i}-\frac{1}{N_i}\right)S_{wi}^2\right]$$

Hence by (8.15) we have

$$E\left[\sum_{i\in J}[N_i\bar{y}_{i.} - \frac{1}{n}\sum_{i\in J} N_i\bar{y}_{i.}]^2\right] = (n-1)S_b^2 + \frac{n-1}{N}\left[\sum_{i=1}^{N} N_i^2\left(\frac{1}{n_i}-\frac{1}{N_i}\right)S_{wi}^2\right]$$

Therefore

$$E\left[N^2\left(\frac{1}{n}-\frac{1}{N}\right)\frac{1}{n-1}\sum_{i\in J}[N_i\bar{y}_{i.} - \frac{1}{n}\sum_{i\in J} N_i\bar{y}_{i.}]^2\right] = V(\hat{Y}_{ms})$$

$$-\left[\sum_{i=1}^{N} N_i^2\left(\frac{1}{n_i}-\frac{1}{N_i}\right)S_{wi}^2\right]$$

Since $\left[\frac{N}{n}\sum_{i\in J} N_i^2\left(\frac{1}{n_i}-\frac{1}{N_i}\right)s_{wi}^2\right]$ is unbiased for $\left[\sum_{i=1}^{N} N_i^2\left(\frac{1}{n_i}-\frac{1}{N_i}\right)S_{wi}^2\right]$, we conclude that

$$N^2\left(\frac{1}{n}-\frac{1}{N}\right)\frac{1}{n-1}\sum_{i\in J}[N_i\bar{y}_{i.} - \frac{1}{n}\sum_{i\in J} N_i\bar{y}_{i.}]^2 + \left[\frac{N}{n}\sum_{i\in J} N_i^2\left(\frac{1}{n_i}-\frac{1}{N_i}\right)s_{wi}^2\right] \text{ is}$$

unbiased for $V(\hat{Y}_{ms})$. ■

Remark If all the first stage units have same number of second stage units, say M and the same number of second stages units is sampled from every sampled primary stage unit, then $\hat{Y}_{ms}$ and $V(\hat{Y}_{ms})$ take the following forms: (Here it is assumed that m is the second stage sample size)

(i) $\hat{Y}_{ms} = \frac{NM}{n} \sum_{i \in J} \bar{y}_{i.}$

(ii) $V(\hat{Y}_{ms}) = \frac{N^2 M^2}{n}\left(\frac{1}{m} - \frac{1}{M}\right) S_w^2 + N^2 M^2 \left(\frac{1}{n} - \frac{1}{N}\right) \bar{S}_b^2$

where $S_w^2 = \frac{1}{N} \sum_{i=1}^{N} S_{wi}^2$ and $\bar{S}_b^2 = \frac{1}{N-1} \sum_{i=1}^{N} (\bar{Y}_{i.} - \bar{\bar{Y}})^2, \bar{\bar{Y}} = \frac{\bar{Y}_{..}}{M}$

It is to be noted that $\bar{S}_b^2 = \frac{S_b^2}{M}$ where S_b^2 is as defined in Theorem 8.2.

Optimum values of *n* and *m*

Now we shall find the values of n and m, namely the sample sizes to be used in the first and second stages of sampling, assuming the conditions stated in Remark 8.2. Naturally these values depend on the type of cost function. If travel between primary stage units is not a major component, then the total cost of the survey can be taken as

$$C = c_1 n + c_2 nm \tag{8.17}$$

The above cost function contains two components. The first component is proportional to the number of *psu*'s to be sampled whereas the second component is proportional to the second stage units to be sampled from each sampled *psu*. Under the above set up, it is possible to find the optimum values of n and m for which $V(\hat{Y}_{ms})$ is minimum for a given cost. Towards this, we consider the function

$$L = V(\hat{Y}_{ms}) + \lambda[c_1 n + c_2 nm - C]$$

$$= \frac{N^2 M^2}{n}\left(\frac{1}{m} - \frac{1}{M}\right) S_w^2 + N^2 M^2 \left(\frac{1}{n} - \frac{1}{N}\right) \bar{S}_b^2 + \lambda[c_1 n + c_2 nm - C] \tag{8.18}$$

Differentiating the above function partially with respect to n and equating the derivative to zero we get

$$-\frac{\bar{S}_b^2}{n^2} - \left(\frac{1}{m} - \frac{1}{M}\right)\frac{S_w^2}{n^2} + c_1 + c_2 m = 0$$

$$n^2 = \frac{\bar{S}_b^2 + \left(\frac{1}{m} - \frac{1}{M}\right) S_w^2}{c_1 + c_2 m} \tag{8.19}$$

Again differentiating partially with respect to m and equating the derivative to zero we get $-\frac{S_w^2}{m^2 n^2} + c_2 = 0$

$$n^2 = \frac{S_w^2}{c_2 m^2} \tag{8.20}$$

Combining (8.19) and (8.20) we get

$$\frac{\bar{S}_b^2+\left(\frac{1}{m}-\frac{1}{M}\right)S_w^2}{c_1+c_2m}=\frac{S_w^2}{c_2m^2}$$

$$c_2\left[\bar{S}_b^2+\left(\frac{1}{m}-\frac{1}{M}\right)S_w^2\right]=\frac{S_w^2}{m^2}(c_1+c_2m)$$

$$c_2\bar{S}_b^2+\left(\frac{c_2S_w^2}{m}-\frac{c_2S_w^2}{M}\right)=\frac{S_w^2c_1}{m^2}+\frac{S_w^2c_2}{m}$$

$$\frac{1}{m^2}=\frac{c_2}{c_1}\frac{\bar{S}_b^2-\frac{S_w^2}{M}}{S_w^2}$$

$$m=\frac{S_w\sqrt{\frac{c_1}{c_2}}}{\sqrt{\bar{S}_b^2-\frac{S_w^2}{M}}} \qquad (8.21)$$

This value can be substituted in (8.17) and a best solution for n can be found easily. The expression given in (8.21) indicates that m is directly proportional to S_w. This implies, the number of second stage units to be taken from a primary stage unit should be large if the variability with respect to y is large within primary stage units. Similarly if the cost per secondary unit c_2 is small or the cost per primary units c_1 is large, m should be large.

2. Two-stage Sampling Under Unequal Probability Sampling

The results presented above are applicable for the case of using simple random sampling in both the stages of sampling. In this section, we shall discuss some results which are quite general in nature and can be applied for any sampling design with known inclusion probabilities. Here it is assumed that all the second stage units in the population are labelled using running numbers from 1 to N_0 where $N_0=\sum_{r=1}^{N} N_r$. N_r being the number of secondary units in the rth *psu*.

If a unit i belongs to the rth *psu*, $r=r(i)$, then the unit will be included in the sample s if

1. the rth *psu* will be included in the first sample; and
2. the unit i will be selected in subsampling, provided that case 1 happened.

Denoting the probability of case 1 by π_r^I and the conditional probability in 2 by π_i^{II}, we have the following formula for the overall inclusion probability

$$\pi_i = \pi_r^I \, \pi_i^{II}, i = 1, 2, \ldots, N_0, r = r(i) \tag{8.12}$$

where $N_0 = \sum_{r=1}^{N} N_r$

For example: If simple random sampling is used in both the stages of sampling.

$$\pi_i = \frac{n}{N}\frac{n_r}{N_r}, r = 1, 2, \ldots, N, i = 1, 2, \ldots, N_0; r = r(i) \tag{8.13}$$

Here it is assumed that a sample of n *psu*'s are selected in the first stage and a subsample of size n_r is drawn from the rth sampled *psu*.

Now we shall consider second-order inclusion probabilities in multistage sampling. For two units i and j belonging to rth and sth *psu*, respectively, we may write

$$\pi_{ij} = \pi_{rs}^I \, \pi_{ij}^{II} \, i, j = 1, 2, \ldots, N_0; r = r(i), s = s(j) \tag{8.14}$$

where π_{rs}^I denotes the probability of simultaneous inclusion of the rth and sth *psu* in the first stage, and π_{ij}^{II} denotes the conditional probability of simultaneous inclusion of the units i and j in subsampling, provided that the rth and sth *psu* have been selected in the first stage.

If $r \neq s$, the subsampling concerning the unit is independent from one concerning the unit j. Therefore

$$\pi_{ij}^{II} = \pi_i^{II} \pi_j^{II} \text{ and } \pi_{ij} = \pi_{rs}^I \, \pi_i^{II} \pi_j^{II} \text{ if } r(i) \neq s(j).$$

If $r = s$, then we have $\pi_{rs}^I = \pi_r^I$ and $\pi_{ij} = \pi_r^I \, \pi_{ij}^{II}$ if $r(i) = r(j)$.

Using these first and second order inclusion probabilities, one can easily construct unbiased estimator for the population total.

The total over the rth *psu* will be denoted by (with respect to the variable y) $T_{ry}, r = 1, 2, \ldots, N$. The set of indices belonging to the *psu* selected in the first stage of sampling will be denoted by $J \subset \{1, 2, \ldots, M\}$ and identified with the first stage sample. The sample of second stage units yielded by subsampling in the rth *psu* will be denoted by $s_r, r = 1, 2, \ldots, N$. Consequently the sample of second stage units will be given by $s = \bigcup_{r \in J} s_r$. Under this notation, we have

$$\pi_r^I = P(r \in J) \text{ and } \pi_{ij}^{II} = P(i \in s \mid r(i) \in J).$$

An unbiased estimator of the population total Y is given by

$$\hat{Y} = \sum_{i \in s} \frac{Y_i}{\pi_i} \tag{8.15}$$

The above estimator can also be expressed as

$$\hat{Y} = \sum_{i \in s} \frac{Y_i}{\pi_r^I \pi_i^{II}} = \sum_{r \in J} \frac{\sum_{i \in s_r} \frac{Y_i}{\pi_i^{II}}}{\pi_r^I} = \sum_{r \in J} \frac{\hat{T}_{ry}}{\pi_r^I}$$

where $\hat{T}_{ry}$ is the estimator of T_{ry} based on subsampling. That is,

$$\hat{T}_{ry} = \sum_{i \in s_r} \frac{Y_i}{\pi_i^{II}}, r = 1, 2, \ldots, N \tag{8.16}$$

The following theorem gives the variance of the above unbiased estimator for the population total.

Theorem 8.4 In two stage sampling

$$V(\hat{Y}) = E_I[\hat{Y}_I - Y]^2 + \sum_{r=1}^{N} \frac{E_{II}(\hat{T}_{ry} - T_{ry})}{\pi_r^I} \tag{8.17}$$

where $\hat{Y}_I = \sum_{r \in J} \frac{\hat{T}_{ry}}{\pi_r^I}$ and the expectations E_I and E_{II} refer to the first-stage sampling and sub-sampling respectively.

Proof Fixing the first stage sample J, we have

$$E_{II}(\hat{Y}) = \sum_{r \in J} E_{II}\left[\frac{\hat{T}_{ry}}{\pi_r^I}\right] = \sum_{r \in J} \frac{\hat{T}_{ry}}{\pi_r^I} = \hat{Y}_I$$

Furthermore, since subsampling is carried out in each sampled first stage unit independently, we have

$$E_{II}(\hat{Y} - Y)^2 = E_{II}[\hat{Y} - E_{II}(\hat{Y})]^2 + \left[E_{II}(\hat{Y} - Y)\right]^2$$

$$= E_{II} \sum_{r \in J} \frac{[\hat{T}_{ry} - T_{ry}]^2}{\pi_r^I} + [\hat{Y}_I - Y]^2$$

$$= \sum_{r \in J} \frac{E_{II}[\hat{T}_{ry} - T_{ry}]^2}{(\pi_r^I)^2} + [\hat{Y}_I - Y]^2$$

Applying the well-known relation $E[.] = E_I E_{II}[.]$ to both the sides of the above expression, we get the required result. ■

Exercises

8.1 Suggest an unbiased estimator for the population total assuming simple random sampling is used in the first stage and *ppswr* sampling is used in the second stage and derive its variance.

8.2 Suppose a population consists of N primary stage units out of which n are selected so that the probability that a sample s of size n is selected, is proportional to sample total of size variable of the primary stage units. Suppose further that for the ith *psu* there is an estimator T_i (based on sampling at the second stage and subsequent stages) of the total Y_i of the primary stage unit. Suggest an unbiased estimator of the population total using T_i's and derive its variance.

8.3 In a two-stage design one subunit is selected with *pp* to x from the entire population. If this happens to come from the *i*th *psu*, a without replacement random sample of $m_i - 1$ subunits is taken from the $M_i - 1$ that remain in the *psu*. From the other $N-1$ *psu*'s a without replacement random sample of $N-1$ *psu*'s is taken. Subsampling of the selected *psu*'s is without replacement simple random. Show that $\dfrac{\sum_{i=1}^{n} M_i \bar{y}_i}{\sum_{i=1}^{n} M_i \bar{x}_i}$ is an unbiased estimator of $\dfrac{Y}{X}$.

Chapter 9

Non-sampling Errors

9.1 Incomplete Surveys

In many large scale surveys, data cannot always be obtained from all the sampled units due to various reasons like the selected respondent may not be available at home and even if present may refuse to co-operate with the investigator etc. In such cases the available data returns are incomplete and some times, this kind of incompleteness called Non-response is so large as to completely vitiate the results. In this section some techniques meant for removing biases arising from incomplete data are presented.

Hansen and Hurwitz Technique

Hansen and Hurwitz (1946) suggested a solution for obtaining unbiased estimates in mail surveys in the presence of non-response. In their method, questionnaires are mailed to all the respondents included in a sample and a list of non-respondents is prepared after the deadline is over. Then a subsample is drawn from the set of nonrespondents and a direct interview is conducted with the selected respondents and the necessary information is collected. The parameter concerned are estimated by combining the data obtained from the two parts of the survey.

Assume that the population is divided into two groups, those who will respond at the first attempt belong to the response class, and those who will not respond called non-response class. Let N_1 and N_2 be the number of units in the population that belong to the response class and the non-response class respectively $(N_1 + N_2 = N)$. Let n_1 be the number of units responding in a simple random sample of size n drawn from the population and n_2 be the number of units not responding in the sample. We may regard the sample of n_1 respondents as a simple random sample from the response class anu the sample of n_2 as a simple random sample from the non-response class. Let h_2 denote the size of the subsample from n_2 non-respondents to be interviewed and $f = \dfrac{n_2}{h_2}$. Unbiased estimators of N_1 and N_2 are given by

$$\hat{N}_1 = \frac{Nn_1}{n} \text{ and } \hat{N}_2 = \frac{Nn_2}{n} \tag{9.1}$$

Let $\bar{y}_{h_2}$ denote the mean of h_2 observations in the subsample and

$$\bar{y}_w = \frac{n_1\bar{y}_1 + n_2\bar{y}_{h_2}}{n} \tag{9.2}$$

The following theorem proves the above estimator is unbiased for the population mean and gives its variance.

Theorem 9.1 The estimator $\bar{y}_w = \dfrac{n_1\bar{y}_1 + n_2\bar{y}_{h_2}}{n}$ is unbiased for the population mean and its variance is

$$V(\bar{y}_w) = \left(\frac{1}{n} - \frac{1}{N}\right)S^2 + (f-1)\frac{N_2}{N}\frac{S_2^2}{n}$$

where S_2^2 is the analogue of S^2 based on the non-response class.

Proof $E(\bar{y}_w) = EE[\bar{y}_w \mid n_1, n_2]$

$$= E\left[\frac{n_1\bar{y}_1 + n_2\bar{y}_{h_2}}{n} \mid n_1, n_2\right]$$

$= E[\bar{y}]$ (since $\bar{y}_{h_2}$ is unbiased for the mean of non-response class)

$= \bar{Y}$

Therefore the estimator is unbiased.

$$V(\bar{y}_w) = VE[\bar{y}_w \mid n_1, n_2] + EV[\bar{y}_w \mid n_1, n_2]$$

$$= V[\bar{y}] + EV[\bar{y}_w \mid n_1, n_2] \tag{9.3}$$

Note that $V[\bar{y}_w \mid n_1, n_2] = V\left[\dfrac{n_1\bar{y}_1 + n_2\bar{y}_{h_2}}{n} \mid n_1, n_2\right]$

$$= V\left[\frac{n_1\bar{y}_1}{n} \mid n_1, n_2\right] + V\left[\frac{n_2}{n}\bar{y}_{h_2} \mid n_1, n_2\right]$$

$$= 0 + \frac{n_2^2}{n^2}\left(\frac{1}{h_2} - \frac{1}{n_2}\right)s_2^2$$

$$= \frac{n_2}{n^2}(f-1)s_2^2$$

where s_2^2 is the sample analogue of S_2^2.

Hence $EV[\bar{y}_w \mid n_1, n_2] = (f-1)E\left[\dfrac{n_2}{n^2}s_2^2 \mid n_2\right] = \dfrac{(f-1)}{n}S_2^2 E\left[\dfrac{n_2}{n}\right]$

$$= \frac{(f-1)}{n}S_2^2\frac{N_2}{N} \tag{9.4}$$

Note that $V(\bar{y}_n)=\left(\frac{1}{n}-\frac{1}{N}\right)S^2$ (9.5)

Substituting (9.4) and (9.5) in (9.3) we get the required result. ■

The cost involved in the above technique contains three components (i) the overhead cost C_0, (2) the cost of collecting, processing per unit in the response class C_1 and (3) the cost of interviewing and processing information per unit in the non-response class C_2. Thus, it is reasonable to consider a cost function of the form $C = C_0 n + C_1 n_1 + C_2 n_2$. Since n_1 and n_2 are random quantities with expectations $n\frac{N_1}{N}$ and $n\frac{N_2}{Nf}$ respectively, the average cost function is $C' = \frac{n}{N}[C_0 N + C_1 N_1 + C_2 \frac{n_2}{f}]$. The following theorem gives the optimum values of f and n for a given variance.

Theorem 9.2 The values of f and n for which the average cost is minimum when $V[\bar{y}_w] = V_0$ are given by
$$f = \frac{C_2\left(S^2 - \frac{N_2 S_2^2}{N}\right)}{S_2^2\left(C_0 + \frac{N_1 C_1}{N}\right)}$$ and
$$n = \frac{S^2 + \frac{N_2(f-1)S_2^2}{N}}{\left(V + \frac{S^2}{N}\right)}$$

Proof of this theorem is left as an exercise.

The problem of incomplete surveys has received the attention of many including El-Bardy(1956), Delenius (1955), Kish and Hess (1959), Bartholomew(1961) and Srinath(1971).

Deming's Model of the effects of call-backs

Deming (1953) developed a mathematical model to study in detail the consequences of different call-back policies. Here the population is divided into r classes according to the probability that the respondent will be found at home. Let w_{ij} = probability that respondent in the jth class will be reached on or before the ith call, p_i = proportion of population falling in the jth class, $\bar{Y}_j$ = mean for the jth class and σ_j^2 = variance for the jth class. Here it is assumed that w_{ij} is positive for all classes. If $\bar{y}_{ij}$ is the mean for those in class j, who were reached on or before the ith call, it is assumed that $E[\bar{y}_{ij}] = \bar{Y}_j$. The true population mean for the item is $\bar{Y} = \sum_{j=1}^{r} [p_j \bar{Y}_j]$.

Suppose a simple random sample of size n is drawn. After i calls, the sample is divided into $(r+1)$ classes: in the first class and interviewed: in the second and interviewed; and so on. The $(r+1)$st class consists of all those not interviewed yet. The numbers falling in these $(r+1)$ classes are distributed according to the multinomial

$$[w_{i1}p_1 + w_{i2}p_2 + \ldots + w_{ir}p_r + (1 - \sum_{j=1}^{r} w_{ij}p_j)]^{n_0}$$

where n_0 is the initial size of the sample. Therefore n_i follows Binomial distribution with parameters n_0 and $\sum_{j=1}^{r} w_{ij}p_j$. For fixed n_i, the number of interviews n_{ij} $(j = 1, 2, \ldots, r)$ follows multinomial with probabilities $\dfrac{w_{ij}p_j}{\sum_{j=1}^{r} w_{ij}p_j}$.

Therefore $E[n_{ij} \mid n_i] = \dfrac{n_i w_{ij} p_j}{\sum_{j=1}^{r} w_{ij} p_j}$

If $\bar{y}_i$ is the sample mean obtained after i calls.

$$E[\bar{y}_i \mid n_i] = E\left[\sum_{j=1}^{r} \frac{n_{ij}\bar{y}_{ij}}{n_i} \mid n_i\right] = \sum_{j=1}^{r} \frac{n_i w_{ij} p_j \bar{Y}_j}{n_i \sum_{j=1}^{r} w_{ij} p_j}$$

$$= \frac{\sum_{j=1}^{r} w_{ij} p_j \bar{Y}_j}{\sum_{j=1}^{r} w_{ij} p_j} = \bar{Y}_i$$

Since the above expected value does not depend on n_i, the overall expectation of $\bar{y}_i$ is also $\bar{Y}_i$. Therefore the estimator is biased for the population mean $\bar{Y}$. The bias of the estimator $\bar{y}_i$ is given by

$$E[\bar{y}_i] = \bar{Y}_i - \bar{Y}$$

$$= \frac{\sum_{j=1}^{r} w_{ij} p_j \overline{Y}_j}{\sum_{j=1}^{r} w_{ij} p_j} - \sum_{j=1}^{r} w_{ij} p_j \overline{Y}_{ij} - \left[1 - \sum_{j=1}^{r} w_{ij} p_j\right] \overline{Y}_i'$$

$$= \frac{\left[1 - \sum_{j=1}^{r} w_{ij} p_j\right]}{\sum_{j=1}^{r} w_{ij} p_j} \sum_{j=1}^{r} w_{ij} p_j (\overline{Y}_{ij} - \overline{Y}_i')$$

where $\overline{Y}_i'$ is the mean of the units not interviewed yet.

The conditional variance of the estimator $\overline{y}_i$ after i calls is

$$V[\overline{y}_i \mid n_i] = \frac{N_i - n_i}{N_i - 1} \frac{1}{n_i} \sum_{j=1}^{r} \left[\frac{N_{ij}}{N_i} [\overline{Y}_{ij} - \overline{Y}_i]^2 + \frac{N_{ik} - 1}{N_{ij}} S_{ij}^2 \right]$$

where $S_{ij}^2 = \frac{1}{N_{ij} - 1} \sum_{k=1}^{N_{ij}} [Y_{ijk} - \overline{Y}_{ij}]^2$. The quantities N_{ij}, N_i etc. have the usual meaning. Taking $a_{ij} = \frac{N_{ij}}{N_i}$ and $\frac{N_{ij} - 1}{N_{ij}} S_{ij}^2 = \sigma_{ij}^2$, $V[\overline{y}_i \mid n_i]$ can be written as

$$V[\overline{y}_i \mid n_i] = \frac{N_i - n_i}{N_i - 1} \frac{1}{n_i} \sum_{j=1}^{r} \{a_{ij} [\overline{Y}_{ij} - \overline{Y}_i]^2 + \sigma_{ij}^2\}$$

If we further assume $\sigma_{ij}^2 = \sigma_j^2$ and ignore terms of order $\frac{1}{n^2}$, it can be seen that

$$MSE[\overline{y}_i] = \left(\frac{1}{n} - \frac{1}{N}\right) \frac{\sum_{j=1}^{r} \{a_{ij} [\overline{Y}_{ij} - \overline{Y}_i]^2 + \sigma_j^2\}}{\sum_{j=1}^{r} w_{ij} p_j}$$

Deming has also considered the problem of determining optimum number of call-backs for the given sample size and cost of the survey. For related results one can refer to Deming (1946).

Politz-Simmons Technique

Politz and Simmons (1949,1950) developed a technique to reduce the bias due to incomplete surveys without successive call-backs. Their method is described below:

The interviewer makes only one call during a specific time on six weekdays. If the respondent is at home, the required information is collected and he is asked how many times in the preceding five days he was at home at the time of visit. This data is used to estimate the probability of the respondent's availability. If the respondent states that he was at home t nights out of five, the ratio $\frac{r+1}{6}$ is taken as an estimator of the frequency π with which he is at home during interviewing hours.

The results from the first call are sorted into six groups according to the values of t (0,1,2,3,4,5). Let n_t be the number of interviews obtained from the ith group and $\bar{y}_t$ the mean based on them. The Politz-Simmons estimate of the population mean is $\hat{\bar{Y}}_{ps} = \dfrac{\sum_{t=0}^{5} \frac{6n_t \bar{y}_t}{t+1}}{\sum_{t=0}^{5} \frac{6n_t}{t+1}}$. In this approach, the fact that the first call results are unduly weighted with persons who are at home most of the time is recognised. Since a person who is at home, on the average, a proportion π of the time has a relative chance π of appearing in the sample, his response should receive a weight $\frac{1}{\pi}$. The quantity $\frac{6}{t+1}$ is used as an estimate of $\frac{1}{\pi}$. Thus $\hat{\bar{Y}}_{ps}$ is less biased than the sample mean from the first call, but it has greater variance because the estimator happens to be weighted mean.

Let the population be divided into classes, people in the jth class being at home π_j of the time. Note that the kth group will contain persons from various classes. That is, persons at home t nights out of the preceding five belong to various classes. Let $n_{jt}, \bar{y}_{jt}$ be the number and the mean for those in class j and group t. Then the estimator $\hat{\bar{Y}}_{ps}$ can be written as $\hat{\bar{Y}}_{ps} = \dfrac{\sum_{t=0}^{5}\sum_{j=1}^{r} \frac{6n_{jt}\bar{y}_{jt}}{t+1}}{\sum_{t=0}^{5}\sum_{j=1}^{r} \frac{6n_{jt}}{t+1}}$. If n_0 is the initial size of sample (response plus not-at-homes) and n_j is the number from class j who are interviewed, the following assumptions are made.

(1) $\frac{n_j}{n_0}$ is a binomial estimate of $p_j \pi_j$

(2) $E[n_{jt} \mid n_j] = n_j \binom{5}{t} \pi_j^t (1-\pi_j)^{5-t}$

(3) $E[\bar{y}_{jt}] = \bar{Y}_j$ for any j and t

It can be shown that under the above assumptions

$$E[\sum_{t=1}^{5} \frac{6}{t+1}] = \frac{n_j}{\pi_j}[1-(1-\pi_j)^6]$$

$$E[\hat{D}] = \sum_{j=1}^{r} \frac{E(n_j)}{\pi_j}[1-(1-\pi_j)^6]$$

$$= n_0 \sum_{j=1}^{r} p_j[1-(1-\pi_j)^6]$$

Since $E[\overline{y}_{jt}] = \overline{Y}_j$ for any j and t, we have

$$E[\hat{\overline{Y}}_{ps}] = \frac{\sum_{j=1}^{r} p_j \overline{Y}_j [1-(1-\pi_j)^6]}{\sum_{j=1}^{r} p_j [1-(1-\pi_j)^6]}$$

This shows that the estimator $\hat{\overline{Y}}_{ps}$ is biased for $\overline{Y}$. However, in practice the amount bias is likely to be small when compared to call-back surveys. The variance of the estimator is quite complicated. For more details, one can refer to the original paper.

9.2 Randomised Response Methods

In many sample surveys involving human populations, it is very difficult to get answers which are truthful and in some cases the respondents fail to co-operate. It is mainly due to sensitivity of certain questions which are likely to affect the privacy of respondents. To overcome this limitation, Warner (1965) has designed a technique to encourage co-operation and truthful answering.

Suppose members of a group A in a population have a socially unacceptable character and we are interested in estimating the proportion π_A of the persons belonging to A. Assume that a simple random sample of size n is drawn with replacement from the given population.

Warner's Method

Each selected respondent is given a random device which results in one of the two statements "I belong to group A " and "I do not belong to A ". The respondent is asked to conduct the experiment unobserved by the investigator and report only "yes" or "no" according to the outcome of the experiment . He does not report the outcome of the experiment. If n_1 persons in the sample report "yes" answer and $n_2 = n - n_1$ report "no" answer then an unbiased estimator of θ, the probability of "yes" answer is given by $\theta_w = \frac{n_1}{n}$. It is to be noted that

$$\theta_W = P\pi_A + (1-P)(1-\pi_A) \tag{9.6}$$

where P is the probability of getting the statement "I belong to group A" and $(1-P)$ is the probability of getting the other statement. It is assumed that P is known. Hence an unbiased estimator of the parameter π_A is $\pi_{AW} = \dfrac{\hat{\theta}_w - (1-P)}{2P-1}, P \neq \dfrac{1}{2}$. Since n_1 has binomial distribution with parameters n and θ_w,

$$V(\pi_{AW}) = \frac{\dfrac{\theta_w(1-\theta_w)}{n}}{(2P-1)^2} \tag{9.7}$$

$$= \frac{\pi_A(1-\pi_A)}{n} + \frac{P(1-P)}{n(2P-1)^2} \tag{9.8}$$

Here it is assumed that the respondent is truthful.

The first term, on the right hand side of (9.8) is the usual binomial variance that would be obtained when all the respondents are willing to answer truthfully and a direct question is presented to each respondent included in the sample. The second term represents a sizable addition due to the random device. It is to be noted that the above method is not useful when $P = \dfrac{1}{2}$ and for $P = 1$, the method reduces to direct questioning.

Simmons Randomised Response Model

In order to enhance the confidence of the respondent in the anonymity provided by the randomised response method, Simmons suggested that one of the statements referred to is a non-sensitive attribute, say Y, unrelated to the sensitive attribute A. In some cases the respondent would get one of the following two statements with probabilities P and $(1-P)$.

1. I belong to group Y
2. I belong to group A

In this case the statement 1 would not embarrass the respondent. If π_Y is the proportion in the population with the attribute Y and it is known then the proportion π_A can be estimated unbiasedly. Note that the probability of getting the yes answer is $\theta_S = P\pi_A + (1-P)\pi_Y$. If $\hat{\theta}_S$ is the proportion of yes answers in the sample of size n, then an unbiased estimator of π_A is

$$\hat{\pi}_{AS} = \frac{\theta_S - (1-P)\pi_Y}{P}$$

and its variance is

$$V(\hat{\pi}_{AS}) = \frac{\theta_S(1-\theta_S)}{nP^2}$$

When π_Y is unknown, the method can be altered to facilitate estimation of both π_Y and π_A. Here, the sample is drawn in the form of two independent samples of sizes n_1 and n_2 again with replacement and with probabilities P_1 and P_2 for getting the sensitive statements in the first and second samples respectively. The same unrelated question is presented in with probabilities $1-P_1$ and $1-P_2$ in the first and second samples respectively. If θ_1 and θ_2 are the respective probabilities of "yes" answer, then we have

$$\theta_1 = P_1\pi_A + (1-P_1)\pi_Y$$

$$\theta_2 = P_2\pi_A + (1-P_2)\pi_Y$$

Solving these two expressions, we get

$$\pi_Y = \frac{P_2\theta_1 - P_1\theta_2}{P_2 - P_1}$$

$$\pi_A = \frac{(1-P_2)\theta_1 - (1-P_1)\theta_2}{P_1 - P_2}$$

Let n'_1 and n'_2 be the number of yes answers in the first and second samples respectively. Since $\frac{n'_1}{n_1}$ and $\frac{n'_2}{n_2}$ are unbiased for θ_1 and θ_2 respectively, an unbiased estimator of π_A is given by

$$\hat{\pi}_{AS} = \frac{(1-P_2)\hat{\theta}_1 - (1-P_1)\hat{\theta}_2}{P_1 - P_2}$$

where $\hat{\theta}_1 = \frac{n'_1}{n_1}$ and $\hat{\theta}_2 = \frac{n'_2}{n_2}$. Since n'_1 and n'_2 are independent and binomially distributed with parameters (n_1, θ_1) and (n_2, θ_2), the variance of $\hat{\pi}_{AS}$ is found to be

$$V(\hat{\pi}_{AS}) = \frac{\frac{(1-P_2)^2\theta_1(1-\theta_1)}{n_1} + \frac{(1-P_1)^2\theta_2(1-\theta_2)}{n_2}}{(P_1 - P_2)^2}$$

Folsom's Model with two unrelated characteristics

Folsom et al (1973) developed an unrelated-question model with two non-sensitive characteristics, y_1 and y_2 in addition to the sensitive character A. Assume that the non-sensitive proportions π_{y1} and π_{y2} are unknown. Two independent simple random samples with replacement of sizes n_1 and n_2 are drawn. Each respondent in both the samples answer a direct question on a non-sensitive topic and also one of two questions selected by a randomised device. The following table given in the next page describes the scheme.

Technique used with respondents	Sample 1	Sample 2
Randomised Response (RR)	Question A	Question A
	Question Y_1	Question Y_2
Direct Response (DR)	Question Y_2	Question Y_1

In both samples let the sensitive question be asked with the probability P, and for $i = 1, 2$, $\lambda_i^r(\lambda_i^d)$ be the probability of a "yes" answer to the question selected by $RR(DR)$ in the ith sample. Then

$$\lambda_1^r = P\pi_A + (1-P)\pi_{Y1} \tag{9.9}$$

$$\lambda_2^r = P\pi_A + (1-P)\pi_{Y2} \tag{9.10}$$

$$\lambda_1^d = \pi_{Y1} \tag{9.11}$$

$$\lambda_2^d = \pi_{Y2} \tag{9.12}$$

Let $\hat{\lambda}_1^r, \hat{\lambda}_2^r, \hat{\lambda}_1^d$ and $\hat{\lambda}_2^d$ denote the usual unbiased estimators of $\lambda_1^r, \lambda_2^r, \lambda_1^d$ and λ_2^d respectively, given by the corresponding sample proportions. Then from (9.9) and (9.12) we get an unbiased estimator as

$$\hat{\pi}_A(1) = \frac{\hat{\lambda}_1^r - (1-P)\hat{\lambda}_2^d}{P} \tag{9.13}$$

Using (9.10) and (9.11) we get another unbiased estimator as

$$\hat{\pi}_A(2) = \frac{\hat{\lambda}_2^r - (1-P)\hat{\lambda}_1^d}{P} \tag{9.14}$$

Variances of the estimators defined in (9.13) and (9.14) can be obtained easily. This is left as an exercise. In addition to these three Randomised Response methods, several other schemes are available. For details one can refer to Chaudhuri and Mukerjee (1988).

9.3 Observational Errors

So far in all our discussions, it has been assumed that each unit in the population us attached a fixed value known as the true value of the unit with respect to the character under study and whenever a population is included in the sample, its value of y is observed. However, this assumption is an over simplification of the problem and actual experience does not support this assumption. There are plenty of examples to show that error of measurements of responses are present when a survey is carried out. In this section we shall consider this problem and devise methods for the measurement of these errors to plan the survey as meticulously as possible.

Let us assume that M interviewers are available for the survey. The response x_{ijk} obtained by interviewer on unit j assumed to be a random variable with $E_2[x_{ijk}] = \overline{X}_{ij}$ and $V_2[x_{ijk}] = S_{ij}^2$. The average of responses obtained by interviewer i on all the N units in the population is $\bar{X}_i = \sum_{j=1}^{N} \frac{X_{ij}}{N}$ and the average obtained by all the M interviewers would be $\bar{X} = \sum_{i=1}^{M} \frac{\bar{X}_i}{M}$. This value can be taken as the expected value of the survey, whereas the true value is $\bar{Y}$, the population mean based on all the units in the population. The difference $\bar{X} - \bar{Y}$ is called the response bias.

The response obtained from a sampled unit depends on the person who observes the unit. Therefore it is desirable to allocate the sample interviewer (selected out of the M available) to the sample units (selected out of the N units in the population). Now consider the situation, in which a simple random sample of $\bar{n} = \frac{n}{m}$ units is selected from the population of N units and assigned to an interviewer selected at random from the population of N units and assigned to an interviewer selected at random from the M available for the study. Another independent sample of size $\bar{n}$ is selected and assigned to another interviewer selected at random from the M. In this process m such subsamples of size $\bar{n}$ are selected and assigned to the M interviewers. The following theorem gives an unbiased estimator of $\bar{X}$ under the above scheme.

Theorem 9.3 Under the sampling scheme described above, an unbiased estimator of $\bar{X}$ is given by $\hat{\bar{X}} = \frac{1}{m}\sum_{i=1}^{m} \bar{x}_i$ where $\bar{x}_i = \frac{1}{\bar{n}}\sum_{i=1}^{\bar{n}} \frac{x_{ijk}}{\bar{n}}$ is the sample mean provided by the ith selection of the interviewer.

Proof If a unit is selected at random from the population containing N units and an interviewer is chosen at random from the M and assigned to the selected unit, the expected value of the response x_{ijk} will be $\bar{X}$. It is because for a given interviewer i and for a given unit j, $E_2[x_{ijk}] = \bar{X}_{ij}$. This implies for a fixed i, $E_2[x_{ijk}] = \frac{1}{N}\sum_{j=1}^{n} \bar{X}_{ij}$. Therefore $E[x_{ijk}] = \frac{1}{MN}\sum_{j=1}^{N} \bar{X}_{ij} = \bar{X}$. This implies that the sample mean $\bar{x}_i$ is unbiased for $\bar{X}$. Hence $E[\bar{x}] = \frac{1}{m}\sum_{i=1}^{m} E[\bar{x}_i] = \bar{X}$. Hence the proof. ■

Theorem 9.4

$$V(\bar{x}) = \frac{V(x)}{n} + \left[\frac{1}{m} - \frac{1}{n}\right]C \text{ where } V[x] = \frac{1}{MN}\sum_{i=1}^{M}\sum_{j=1}^{N} E[x_{ijk} - \bar{X}]^2$$

$$\text{and } C = \frac{1}{MN(N-1)}\sum_{i=1}^{M}\sum_{j<j'}^{N} E[x_{ijk} - \bar{X}][x_{ij'k} - \bar{X}]$$

Proof of this theorem follows from routine algebra and hence left as an exercise.

The variance of the sample mean based on a survey employing interviewers has two components. One is the variability of all responses over all units to all interviewers and the other is the covariance between responses obtained from different units within interviewer assignments. If advance estimates of these two components are available, one can determine from the variance given in Theorem 9.4 the optimum number of interviewers to employ for the collection of data.

Let c_1 be the cost per unit in the sample and c_2 be the cost per interviewer, so that the total cost of the surveyor is

$$c_t = c_0 + c_1 n + c_2 m \tag{9.15}$$

The values of n and m can be found by minimising $V(\bar{x})$ for a given cost with the help of the method of Lagrangian multipliers. Setting the partial derivatives of $V(\bar{x}) + \lambda\ (c_0 + c_1 n + c_2 m - c_t)$ with respect to n and m equal to zero, we get

$$\lambda c_1 = \frac{V(x) - C}{n^2} \text{ and } \lambda c_2 = \frac{C}{m^2} \tag{9.16}$$

$$\frac{m}{n} = \frac{\sqrt{\frac{c_1}{c_2}}}{\sqrt{V(x) - C}}\sqrt{C} \tag{9.17}$$

The actual values of n and m are obtained by substituting the ratio given in (9.17) in the cost function defined in (9.15). Since the covariance component C and the variance V depend on the number of interviewers used and the size of the assignment, the solution obtained should be used for getting an idea of the magnitudes involved. Thus we have seen the manner in which resources can be allocated towards the reduction of sampling errors (as provided by n) and non-sampling errors (interviewer errors). The following theorem gives unbiased estimates of $C, V(x)$ and $V(\bar{x})$ under the sampling scheme described in this section.

Theorem 9.5 Under the sampling scheme described in this section, unbiased estimates of $C, V(x)$ and $V(\bar{x})$ are given by $C = \dfrac{s_b^2 - s_w^2}{\bar{n}}, V(x) = s_w^2 + \dfrac{s_b^2 - s_w^2}{\bar{n}}$

and $V(\bar{x}) = \frac{1}{m(m-1)} \sum_{i=1}^{m} [\bar{x}_i - \bar{x}]^2$ where $s_b^2 = \frac{1}{m-1} \sum_{i=1}^{m} [\bar{n}(\bar{x}_i - \bar{x})]^2$ and

$$s_w^2 = \frac{1}{m(\bar{n}-1)} \sum_{i=1}^{m} \sum_{j=1}^{\bar{n}} [x_{ijk} - \bar{x}_i]^2$$

Proof of this theorem is left as an exercise.

Exercises

9.1 Extend Warner's method to the case of estimating two proportions.

9.2 Find the mean square error of the estimator $\pi_{ab} = a\theta + b$ where θ is as defined in Warner's model.

9.3 Find the minimum mean square error of the estimator suggested in 9.2 and offer your comments.

9.4 Obtain an unbiased estimator for the sensitive proportion under Warner's method assuming, the probability of a respondent being untruthful is L and derive the variance of the estimator.

Chapter 10

Recent Developments

10.1 Adaptive Sampling

It has been an untiring endeavour of researchers in sampling theory to seek estimators with increased precision. In the earlier chapters we have seen a variety of sampling estimating strategies which use the information of a suitable auxiliary (size) variable either in the sampling design or in the estimator. There are very few sampling schemes which use the knowledge of study variable in the selection stage. Recently Thompson (1990) introduced sampling schemes which directly use the knowledge of study variable in the selection process. In this section details of his sampling schemes are presented. Quite often we encounter surveys where the investigator gathers information regarding the number of individuals having some specific characteristics. As an example one can think of a survey involving endangered species in which observers record data regarding the number of individuals of the species seen or heard at locations within a study area. In such surveys frequently zero abundance is encountered. In those cases, whenever substantial abundance is seen, exploration in nearby locations is likely to yield additional clusters of abundance. These kinds of patterns are encountered along with others, from whales to insects, from trees to lichens and so on.

Generally, in sample surveys, survey practitioners decide their sampling strategy before they actually begin data collection. However, functioning in this predetermined manner may not be effective always. For example, in epidemiological studies of contagious diseases, whenever an infected individual is encountered, it is highly likely that neighbouring individuals will reveal a higher than expected incidence rate. In such situations, field workers may not like to stick to their original sampling plan. They will be interested in departing from the preselected sample plan and add nearby or closely associated units to the sample. Keeping these points in mind, Thompson (1990) suggested a new sampling scheme. In the sampling scheme suggested by him namely, "**adaptive sampling**", an initial sample of predetermined size is drawn according to a conventional sampling design. The values of sampled units are scrutinized. Whenever the observed value of a selected unit satisfies a given condition of interest, additional units are added to the sample from the neighbourhood of that unit. The basic idea of the design is illustrated in Figures 10a and 10b. Figure

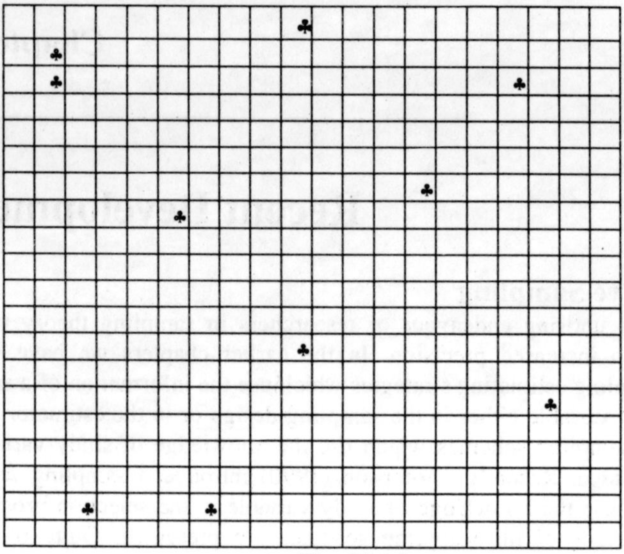

Fig. 10 (a)
Initial sample of 10 units

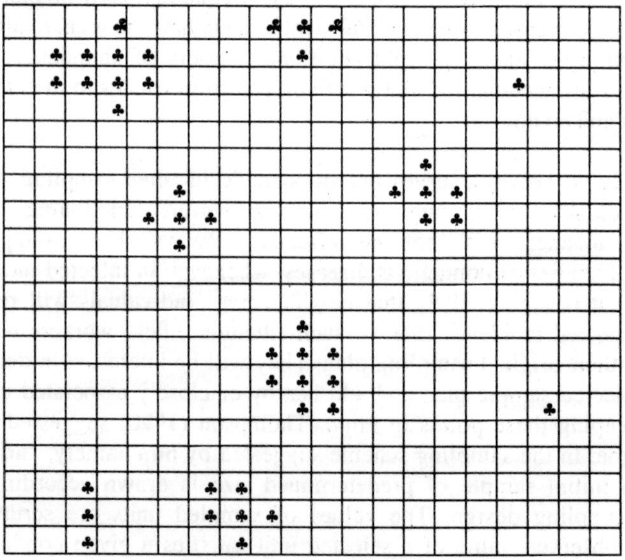

Fig. 10 (b)
Final sample after neighboring units are included

10a shows an initial sample of 10 units. Whenever one or more of the units is found to satisfy a given condition, the adjacent neighbouring units to the left, right, top and bottom are added to the sample. When this process is completed, the sample consists of 45 units, shown in Figure 10b. It is pertinent to note that neighbourhood of units may be defined in so many ways other than spatial proximity. The formal definition of adaptive sampling is presented below:

In adaptive sampling an initial set of units is selected by some probability sampling procedure, and whenever variable of interest of a selected unit satisfies the given criterion, additional units in the neighbourhood of that unit are added to the sample. The criterion for selection of additional neighbouring units can be framed in several ways, depending on the nature of study. For example, the criterion for additional selection of neighbouring units can be taken as an interval or a set C which contains a given range of values with respect to the variable of interest. The unit i is said to satisfy the condition if $Y_i \in C$. For example, a unit satisfies the condition if the variable of interest Y_i is greater than or equal to some constant c. That is, $C = \{x : x \leq c\}$. Here it is assumed that the initial sample consists of simple random sample of size n units selected either with or without replacement. To introduce appropriate estimators under adaptive sampling scheme, we need the following definitions.

Neighbourhood of a unit For any unit in the population, the neighbourhood of a unit U_i is defined as collection of units which includes unit U_i with the property that if unit U_j is in the neighbourhood of unit U_i, then the unit U_i is in the neighbourhood of unit U_j. These neighbourhoods do not depend on the population values.

Cluster The collection of all units that are observed under the design as a result of initial selection of unit U_i is termed as cluster. Note that such a collection may consist of the union of several neighbourhoods.

Network A set of units is known as a network if selection in the initial sample of any unit in the set will result in inclusion in the final sample of all units in that network.

It is convenient to consider any unit not satisfying the condition a network of size one, so that the given y-values may be uniquely partitioned into networks.

Edge unit A population unit is said to be edge unit if it does not satisfy the condition but is in the in the neighbourhood of one that satisfies the condition.

Notations n_1 : Size of initial sample

ψ_k : Network which consists of the unit U_k

m_k : Number of units in the network to which unit k belongs

a_i : Total number of units in networks of which unit i is an edge unit.

Selection procedure and related properties

As mentioned earlier, an initial sample consisting of n_1 units using *SRSWR* or *SRSWOR* is sampled. When a selected unit satisfies the condition all units within its neighbourhood are added to the sample and observed. Note that in addition to units satisfying the condition, even those units in their neighbourhoods are also included in the sample and so on.

Denote by m_i the number of units in the network to which unit i belongs and by a_i the total number of units in networks of which unit i is an edge unit. Note that if unit i satisfies the criterion C then a_i =0, whereas if unit i does not satisfy the condition then $m_i = 1$. It may be noted that the unit i will be selected in a given draw if either any one of the m_i units in its network is drawn in the initial sample or any one of the a_i units for which this is an edge unit, is drawn in the sample. Hence the probability of selection for the unit in a given draw is $p_i = \frac{m_i + n_i}{N}$. The number of ways of choosing n_1 units out of N is $\binom{N}{n_1}$. Let B_i be the subset of the population units containing either the units which are in the network containing the unit i or the units for which unit i is an edge unit. Clearly $n(B_i) = m_i + a_i$. A sample not containing the unit i can be drawn by considering the set $S - B_i$ which contains $N - m_i - a_i$ units. Hence the probability of not including unit i in the sample is $\frac{\binom{N - m_i - a_i}{n_1}}{\binom{N}{n_1}}$. Therefore the probability of including the unit i in the sample is $\alpha_i = 1 - \frac{\binom{N - m_i - a_i}{n_1}}{\binom{N}{n_1}}$. When the initial sample is selected by using *SRSWR*, the probability that the unit i is included in the sample is $\alpha_i = 1 - (1 - p_i)^{n_i}$. Since some of the a_i may not be known, the draw by draw probability p_i as well as the inclusion probability α_i cannot be determined.

Estimators under Adaptive sampling

Classical estimators such as sample mean are not unbiased under adaptive sampling. Now we shall describe some estimators suitable for adaptive sampling.

(i) The initial sample size

If the initial sample in the adaptive design is selected either by *SRSWR* or *SRSWOR*, the mean of the initial observations is unbiased for the population mean. However, this estimator completely ignores all observations in the sample other than those initially selected.

(ii) Modified Hansen-Hurwitz estimator

In conventional sampling, Hansen-Hurwitz estimator, in which each y-value is divided by the number of times the unit is selected, is an unbiased estimator of population mean. However, in adaptive sampling , selection probabilities are not known for every unit in the sample. An unbiased estimator can be formed by modifying the Hansen-Hurwitz estimator to make use of observations not satisfying the condition only when they are selected in the sample. Let ψ_k denote the network which consists of the unit U_k and m_k be the number of units in that network. Let $\overline{y}_k^*$ be the average of observations in the network that includes the kth unit of the initial sample. That is $\overline{y}_k^* = \frac{1}{m_k}\sum_{i \in \psi_k} Y_i$. A modified Hansen-Hurwitz estimator can be defined by using $\overline{y}_k^*$ as $\overline{y}_{HH}^* = \frac{1}{n_1}\sum_{i=1}^{n_1}\overline{y}_i^*$

Theorem 10.1 The estimator $\overline{y}_{HH}^* = \frac{1}{n_1}\sum_{i=1}^{n_1}\overline{y}_i^*$ is unbiased for the population mean.

Proof **Case 1** When *SRSWOR* is used to select initial sample.

Let z_i indicates the number of times the ith unit of the population appears in the estimator, The random variable z_i has a hypergeometric distribution when initial sample is selected by *SRSWOR* with $E[z_i] = \frac{n_1 m_i}{N}$.

$$\text{Therefore } \overline{y}_{HH}^* = \frac{1}{n_1}\sum_{k=1}^{n_1}\overline{y}_k^* = \frac{1}{n_1}\sum_{k=1}^{n_1}\frac{1}{m_k}\sum_{j \in \psi_k} Y_j$$

$$= \frac{1}{n_1}\sum_{k=1}^{N}\frac{z_k Y_k}{m_k}$$

Taking expectation on both the sides we get the required result.

Case 2 When *SRSWOR* is used to select initial sample.

Let z_i, as in case 1,indicates the number of times the ith unit of the population appears in the estimator. It is to be noted that z_i is nothing but the number of times the network including the unit i is represented in the sample. Note that $P(z_i) = \binom{n_1}{z_i}\left(\frac{m_i}{N}\right)^{z_i}\left(1 - \frac{m_i}{N}\right)^{n-z_i}$ Therefore $E[z_i] = \frac{n_1 m_i}{N}$. Expressing the estimator $\overline{y}^*_{HH} = \frac{1}{n_1}\sum_{k=1}^{n_1}\overline{y}^*_k$ in the form considered in case 1 and taking expectations we get the required result. ■

The following theorem gives the variance of the estimator $\overline{y}^*_{HH} = \frac{1}{n_1}\sum_{k=1}^{n_1}\overline{y}^*_k$ in the two cases of simple random sampling.

Theorem 10.2 (a) If the initial sample is selected by *SRSWOR*, the variance of $\overline{y}^*_{HH} = \frac{1}{n_1}\sum_{k=1}^{n_1}\overline{y}^*_k$ is given by $\frac{N-n_1}{Nn_1}\frac{1}{N-1}\sum_{i=1}^{N}[\overline{y}^*_i - \mu]^2$ (b) If the initial sample is selected by *SRSWOR*, the variance of $\overline{y}^*_{HH} = \frac{1}{n_1}\sum_{k=1}^{n_1}\overline{y}^*_k$ is given by $\frac{n_1}{N-1}\sum_{i=1}^{N}[\overline{y}^*_i - \mu]^2$,where $\overline{y}^*_i$ is the average of observations in the network that includes the *k*th unit of the initial sample and $\mu = \frac{1}{N}\sum_{i=1}^{N} Y_i$.

Proof Taking $\overline{y}^*_i$ as the variable of interest and applying the results available under non-adaptive sampling scheme, the desired expressions can be obtained. ■

(iii) Modified Horvitz-Thompson Estimator

We know that the knowledge of first order inclusion probabilities π_i can be used to construct the Horvitz-Thompson estimator for estimating the population total . With the adaptive designs, the inclusion probabilities are not known for all units included in the sample. Hence it can not be used to estimate the total unbiasedly. An unbiased estimator can be formed by modifying the Horvitz-Thompson estimator to make use of observations not satisfying the condition only when they are included in the initial sample. In this case, the probability that a unit is used in the estimator can be computed, even though its actual probability of inclusion in the sample may be unknown.

Define the indicator variable

$J_k = 0$ if the *k*th unit in the sample does not satisfy the condition and was not included in the sample

$= 1$ otherwise, for $k = 1, 2, ..., N$.

The modified estimator is $\overline{y}^*_{HT} = \frac{1}{N}\sum_{k=1}^{\nu}\frac{y_k J_k}{\alpha^*_k}$ where ν is the number of distinct units in the sample α^*_k is the probability that *i* is included in the estimator. It can be seen that, whether the unit *i* satisfies the condition or not, the probability of including the unit in the estimator is $1-\frac{\binom{N-m_k}{n_1}}{\binom{N}{n_1}}$. The following theorem gives the variance of the estimator $\bar{y}^*_{HT}$.

Theorem 10.3 The estimator $\bar{y}^*_{HT}$ is unbiased for the population mean and its variance is $\frac{1}{N^2}\sum_{j=1}^{D}\sum_{h=1}^{D} y_h y_j \left[\frac{\pi_{jh}-\pi_{\pi}\pi_j}{\pi_{\pi}\pi_j}\right]$ where D is the number of networks in the population and π_{jh} is the probability that the initial sample contains atleast one unit in each of the networks j and h.

Proof of this theorem is straight forward and hence omitted.

The results presented above are due to Thompson and more about adaptive sampling are available in Thompson (1990,1991a,1991b).

10.2 Estimation of Distribution Function

The problem of estimating finite population totals, means and ratios of the survey variables are widely discussed in sample survey literature. But estimation of finite population distribution function has not received that much attention. Estimation of distribution function is often an important objective because sometimes it is necessary to identify subgroups in the population whose values for particular variables lie below or above the population average, median, quantiles or any other given value. The of distribution function in finite population mean has received the attention of Chambers and Dunstan (1986) and Rao, Kovar and Mantel (1990). In this section their contributions are presented.

Population Distribution Function : Let $\Delta(x)$ be the step function

$$\Delta(x) = 1 \text{ if } x \le 0$$
$$=0 \text{ otherwise}$$

Let $Y_1, Y_2, ..., Y_N$ be the values of the N units in the population with respect to the variable y. The finite population distribution function of y is defined as

$$F_N(t) = \frac{1}{N}\sum_{i=1}^{N} \Delta(t - Y_i), t \in R \tag{10.1}$$

We know that the Horvitz-Thompson estimator for a finite population total Y is given by

$$\hat{Y}_{HT} = \sum_{i \in s} \frac{Y_i}{\pi_i} \tag{10.2}$$

where π_i's are the inclusion probabilities corresponding the sampling design $P(s)$ used to choose the sample. Hence the Horvitz-Thompson estimators of $\sum_{i=1}^{N} 1$ and $\sum_{i=1}^{N} \frac{\Delta(t-Y_i)}{\pi_i}$ are $\sum_{i \in s} \frac{1}{\pi_i}$ and $\sum_{i \in s} \frac{\Delta(t-Y_i)}{\pi_i}$ respectively. Hence a design-based estimator of $F_N(t)$ defined in (10.1) is $\hat{F}_N(t) = \dfrac{\left\{\sum_{i \in s} \frac{\Delta(t-Y_i)}{\pi_i}\right\}}{\left\{\sum_{i \in s} \frac{1}{\pi_i}\right\}}$.

Note that $\hat{F}_N(t)$ reduces to the ordinary sample empirical distribution function and it is design unbiased under any sampling design satisfying $\sum_{i \in s} \frac{1}{\pi_i} = N$.

Kuk(1988) compared the performance of the estimator $\hat{F}_N(t)$ with those of $\hat{F}_L(t)$ and $\hat{F}_R(t)$ where

$$\hat{F}_L(t) = \frac{1}{N} \sum_{i \in s} \Delta \frac{(t-Y_i)}{\pi_i}$$

$$\hat{F}_R(t) = 1 - S_R(t), S_R(t) = \frac{1}{N} \sum_{i \in s} \Delta \frac{(t-Y_i)}{\pi_i}$$

It may be noted that $S_R(t)$ estimates the proportion of units in the population whose values exceed the given value t. It is interesting to note that $\hat{F}_L(t)$ is not necessarily equal to $\hat{F}_R(t)$. Further it can be easily seen that both $\hat{F}_L(t)$ and $\hat{F}_R(t)$ are unbiased for $F_N(t)$ under all sampling designs for which $\pi_i > 0$ for every $i = 1, 2, ..., N$. Even though both of them are unbiased for $F_N(t)$, they lack the most important property of being distribution functions. On the other hand, $\hat{F}_N(t)$ even though by nature a distribution function, it is not unbiased. The following theorem gives the mean square errors of $\hat{F}_L(t)$ and $\hat{F}_R(t)$ and the approximate mean square error of $\hat{F}_N(t)$.

Theorem 10.4 (a) The mean square error of $\hat{F}_L(t)$ is

$$\frac{1}{N^2} \sum_{i=1}^{N} \sum_{j=1}^{N} \frac{\pi_{ij} - \pi_i \pi_j}{\pi_i \pi_j} \Delta(t-Y_i) \Delta(t-Y_j)$$

(b) The mean square error of $\hat{F}_R(t)$ is

$$\frac{1}{N^2} \sum_{i=1}^{N} \sum_{j=1}^{N} \frac{\pi_{ij} - \pi_i \pi_j}{\pi_i \pi_j} \Delta(Y_i - t) \Delta(Y_j - t)$$

(c) The approximate mean square error of $\hat{F}_N(t)$

$$\frac{1}{N^2}\sum_{i=1}^{N}\sum_{j=1}^{N}\frac{\pi_{ij}-\pi_i\pi_j}{\pi_i\pi_j}[\Delta(t-Y_i)-F_N(t)][\Delta(t-Y_j)-F_N(t)]$$

Proof of this theorem is straight forward and hence left as an exercise.

Remarks (1) Further it can be seen that $MSE(\hat{F}_R(t)) \le MSE(\hat{F}_L(t))$ if $\sum_{i=1}^{N} b_i \ge \sum_{i=1}^{N} b_i \Delta(Y_i - t)$ where $b_i = \sum_{j=1}^{N}\frac{\pi_{ij}-\pi_i\pi_j}{\pi_i\pi_j}$.

(2) The results mentioned above are due to Kuk(1988) and more details can be obtained from the original paper.

Rao et al(1990) suggested difference and ratio estimators for population distribution function which use the knowledge of auxiliary information. The design based ratio and difference estimator of $\hat{F}_N(t)$ are obtained from standard results for totals or means treating $\Delta(t-Y_i)$ and $\Delta(t-\hat{R}X_i)$ as y and x variables respectively, where $\hat{R} = \dfrac{\sum_{i\in s}\left[\frac{Y_i}{\pi_i}\right]}{\sum_{i\in s}\left[\frac{X_i}{\pi_i}\right]}$ is the customary design based estimator of the population ratio $R = \dfrac{Y}{X}$. The ratio estimator of the population distribution function is given by

$$\hat{F}_r(t) = \frac{1}{N}\frac{\sum_{i\in s}\left[\frac{\Delta(t-Y_i)}{\pi_i}\right]}{\sum_{i\in s}\left[\frac{\Delta(t-\hat{R}X_i)}{\pi_i}\right]}\sum_{i=1}^{N}\Delta(t-\hat{R}X_i)$$

which reduces to $\hat{F}_N(t)$ when Y_i is proportional to X_i for all i. Hence the variance will be zero if Y_i is proportional to X_i. This suggests that $\hat{F}_r(t)$ could lead to considerable gains in efficiency over $\hat{F}_N(t)$, when Y_i is approximately proportional to X_i. The difference estimator with the same desirable property is given by

$$\hat{F}_d(t) = \frac{1}{N}\left[\sum_{i\in s}\left[\frac{\Delta(t-Y_i)}{\pi_i}\right]+\sum_{i=1}^{N}\Delta(t-\hat{R}X_i)-\sum_{i\in s}\left[\frac{\Delta(t-\hat{R}X_i)}{\pi_i}\right]\right]$$

Using the data of Chambers and Dunstan (1986), the performance of the above two estimators were studied by Rao et al (1990). They found that the difference estimator is less biased than the ratio estimator for smaller values of $F_N(t)$.

They also found that $\hat{F}_d(t)$ is more precise than $\hat{F}_r(t)$ and $\hat{F}_N(t)$. The presence of $\hat{R}$ in $\hat{F}_r(t)$ and $\hat{F}_d(t)$ creates difficulties in evaluating (analytically) the exact bias and the mean square errors of these estimators. Invoking the results of Randles (1982), they obtained the approximate design variances of $\hat{F}_r(t)$ and $\hat{F}_d(t)$ which are given below :

$$V[\hat{F}_d(t)] = \frac{1}{N^2} V[\Delta(t - Y_i) - \Delta(t - RX_i)]$$

$$V[\hat{F}_r(t)] = \frac{1}{N^2} V\left[\Delta(t - Y_i) - \left\{\frac{F_y(t)}{F_x(\frac{t}{R})}\right\}\Delta(t - RX_i)\right]$$

where $V(Y_i) = \sum_{i=1}^{N}\sum_{\substack{j=1 \\ i<j}}^{N} (\pi_{ij} - \pi_i\pi_j)\left[\frac{Y_i}{\pi_i} - \frac{Y_j}{\pi_j}\right]^2$

The estimated variances of $\hat{F}_r(t)$ and $\hat{F}_d(t)$ are

$$v[\hat{F}_d(t)] = \frac{1}{N^2} v[\Delta(t - Y_i) - \Delta(t - \hat{R}X_i)]$$

and $$v[\hat{F}_r(t)] = \frac{1}{N^2} v\left[\Delta(t - Y_i) - \left\{\frac{\hat{F}_y(t)}{\hat{F}_x(\frac{t}{\hat{R}})}\right\}\Delta(t - \hat{R}X_i)\right]$$

where $v(Y_i) = \sum_{i \in s}\sum_{\substack{j \in s \\ i<j}} \left[\frac{Y_i}{\pi_i} - \frac{Y_j}{\pi_j}\right]^2 \left(\frac{\pi_i\pi_j - \pi_{ij}}{\pi_{ij}}\right)$ and $\hat{F}_x(t)$ and $\hat{F}_y(t)$ are the customary estimates of $F_x(t)$ and $F_y(t)$ respectively.

10.3 Randomised Response Method for Quantitative Data

In the last chapter, we have seen several randomised response methods which are meant for estimating the proportion of units in a population possessing a sensitive character. In this section, a randomised response method meant for dealing with quantitative data as developed by Eriksson (1973a,b) is presented. This problem arises when one is interested in estimating the earnings from illegal or clandestine activities, expenses towards gambling or consumption of alchoholic and so on. These are some examples where people prefer not to reveal their exact status. Let $Y_1, Y_2, ..., Y_N$ be the unknown values of N units labelled $i = 1, 2, ..., N$ with respect to the sensitive study variable y. To estimate the population total Y, Eriksson (1973a,b) suggested the following procedure.

A sample of desired size is drawn by using the sampling design $P(s)$. Let $X_1, X_2, ..., X_N$ be predetermined real numbers supposed to cover the anticipated range of unknown population values $Y_1, Y_2, ..., Y_N$. The quantities $q_j, j = 1, 2, ..., M$ are suitably chosen non-negative proper fractions and C is a rightly chosen positive proper fraction such that $C + \sum_{j=1}^{M} q_j = 1$. Each respondent included in the sample is asked to use conduct a random experiment independently $k(>1)$ times each to produce random observations $Z_{ir}, r = 1, 2, ..., k$,

$$\begin{aligned} Z_{ir} &= Y_i \quad \text{with probability C} \\ &= X_j \quad \text{with probability } q_j, j = 1, 2, ..., M \end{aligned}$$

A corresponding device is independently used for every sampled individual so that the values $Z_{ir}, r = 1, 2, ..., k$, for $i \in s$ are generated. For theoretical purpose, the random vectors $\underline{Z}_r = (Z_{1r}, Z_{2r}, ..., Z_{Nr}), r = 1, 2, ..., k$ are supposed to be defined for every unit in the population. Let $\underline{Z} = (\underline{Z}_1, \underline{Z}_2, ..., \underline{Z}_k)$. Denote by E_R, V_R and C_R taking expectation, variance and covariance with respect to the randomisation technique employed to yield Z_{ir} values. Let $\overline{Z}_i = \frac{1}{k}\sum_{r=1}^{k} Z_{ir}$ and

$$\mu_x = \frac{1}{1-C}\sum_{j=1}^{M} q_j X_j .$$

$$\begin{aligned} \text{Note that } E_R[Z_{ir}] &= CY_i + \sum_{j=1}^{M} q_j X_j \\ &= CY_i + (1-C)\mu_x \end{aligned} \tag{10.3}$$

Hence $E_R[\overline{Z}_i] = CY_i + (1-C)\mu_x, i = 1, 2, ..., N; r = 1, 2, ..., k$.

Therefore an estimator of Y_i is given by $\hat{Y}_i = \dfrac{\overline{Z}_i - (1-C)\mu_x}{C}$. (10.4)

A general estimator for the population total and also its variance is given in the theorem furnished below.

Theorem 10.5 An unbiased estimator for the population total is given by $e(s, \underline{Z}) = a_s + \sum_{i \in s} b_{si} Y_i$, where a_s and b_{si} are free of $Y_1, Y_2, ..., Y_N$ and satisfy $\sum_s a_s P(s) = 0$ and $\sum_{s \ni i} b_{si} P(s) = 1$, $i = 1, 2, ..., N$. The variance of $e(s, \underline{Z})$ is given by $V_P[e(s, \underline{Y})] + \frac{1-C}{kC^2}\sum_{i=1}^{N}\sigma_i^2 \sum_s b_{si}^2 P(s)$. Here $\sum_s$ is the sum over all samples.

Proof Taking E_P, V_P and C_P as operators for expectation, variance and covariance with respect to the design. Assuming commutativity, we write $E_{PR} = E_P E_R = E_R E_P = E_{RP}, V_{PR} = V_{RP}$ to indicate operators for expectation and variance respectively, with respect to randomisation followed by sampling, or vice versa. Taking expectation for the estimator $e(s, \underline{Z})$ we get $E_R[e(s,\underline{Z})] = a_s + \sum_{i \in s} b_{si} E_R[Y_i]$

$$= a_s + \sum_{i \in s} b_{si} Y_i$$

Again taking expectation with respect to the sampling design, we note that

$$E_P E_R[e(s,\underline{Z})] = Y$$

The variance of $e = e(s,\underline{Z})$ can be written as

$$V_{PR}(e) = V_P E_R[e] + E_P V_R[e] \tag{10.5}$$

Denoting by $\sigma_{xx}^2 = \frac{1}{1-C} \sum_{j=1}^{M} q_j (X_j - \mu_x)^2$ and $\sigma_i^2 = \sigma_x^2 + C(Y_i - \mu_x)^2$, we write $V_R[Z_{ir}] = (1-C)\ [\sigma_x^2 + C(Y_i - \mu_x)^2]$

$$= (1-C)\ \sigma_i^2,\ i = 1, 2, \ldots, N\ ;\ r = 1, 2, \ldots, k\ ;$$

Therefore $V_R[e] = \frac{1}{kC^2} \sum_{i \in s} b_{si}^2 V_R(Z_{ir})$

$$= \frac{1-C}{kC^2} \sum_{i \in s} b_{si}^2 \sigma_i^2$$

Hence $$V_{PR}[e] = V_P[e(s,\underline{Y})] + \frac{1-C}{kC^2} \sum_{i=1}^{N} \sigma_i^2 \sum_{s \ni i} b_{si}^2 P(s) \tag{10.6}$$

Hence the proof. ■

Note The second term in the right hand side of (10.6) shows how variance increases (efficiency is lost) when one uses randomised response method rather than direct survey.

Under designs yielding positive first order inclusion probabilities for all units and positive second order inclusion probabilities for all pairs of units, an unbiased estimator for the above variance can be found easily in particular when a_s =0 as shown below.

When a_s =0 ,the variance of the estimator with respect to the sampling design can be written as

$$V_P[e(s,\underline{Y})] = \sum_{i=1}^{N} c_i Y_i^2 + \sum_{i=1}^{N} \sum_{\substack{j=1 \\ i \neq j}}^{N} d_{ij} Y_i Y_j$$

Denote by $v(s,\underline{Y}) = \sum_{i \in s} f_{si} Y_i^2 + \sum_{i} \sum_{\substack{j \neq i \\ i,j \in s}} g_{sij} Y_i Y_j$ where f_{si}'s and g_{sij}'s quantities free of $\underline{Y}$ satisfying $E_P[v(s,\underline{Y})] = V(s,\underline{Y})$.

Note that if $\varphi_i = \frac{1}{C}\left[\sum_{r=1}^{k} \frac{Z_{ir}^2}{k} - (1-C)(\sigma_x^2 + \mu_x^2)\right]$ then $E_R(\varphi_i) = Y_i^2$ and $E_R(\hat{Y}_i \hat{Y}_j) = Y_i Y_j, i \neq j \in s$. Therefore $v(s,\underline{Z}) = \sum_{i \in s} f_{si} \varphi_i + \sum_{i} \sum_{\substack{j \neq i \\ i,j \in s}} g_{sij} \hat{Y}_i \hat{Y}_j$

satisfies
$$E_{PR}[v(s,\underline{Z})] = E_P[E_R v(s,\underline{Z})] = E_P[v(s,\underline{Y})] = V_P[e(s,\underline{Y})]$$

Further if $s_{zi}^2 = \frac{1}{k-1}\sum_{r=1}^{k}[Z_{ir} - \overline{Z}_i]^2$, then $E_R[s_{zi}^2] = V_R[Z_{ir}], r = 1, 2, \ldots, k$.

Hence $E_R\left[\frac{1}{kC^2}\sum_{i \in s} b_{si}^2 s_{zi}^2\right] = \frac{1}{kC^2}\sum_{i \in s} b_{si}^2 V_R(Z_{ir}) = V_R(e)$

Taking expectation with respect to the sampling design, we have

$$E_{PR}\left[\frac{1}{kC^2}\sum_{i \in s} b_{si}^2 s_{zi}^2\right] = E_P[V_R(e)]$$

As a result of the above discussion, we have

$$E_{PR}\left[v(s,\underline{Z}) + \frac{1}{kC^2}\sum_{i \in s} b_{si}^2 s_{zi}^2\right] = V_{PR}(e)$$

Therefore $v(s,\underline{Z}) + \frac{1}{kC^2}\sum_{i \in s} b_{si}^2 s_{zi}^2$ is an unbiased estimator for $V_{PR}(e)$.

For more details about randomised response methods, one can refer to the monograph by Chaudhuri and Mukerjee (1988).

References

1. Bartholomew, D.J. (1961) : A method of allowing for "not-at-homes" bias in sample surveys, App. Stat., 10,52-59.
2. Chambers, R.L. and Dunstan, R. (1986) : Estimating distribution function from survey data, Biometrika, 73,3,597-604.
3. Cochran, W.G. (1946) : Relative accuracy of systematic and stratified random samples for a certain class of populations, Ann. Math. Stat., 17, 164-177.
4. Delenius, T. (1955) : The problem of not-at-homes, Statistisk Tidskrift., 4,208-211.
5. Deming, W.E. (1953) : On a probability mechanism to obtain an economic balance between the resulting error of response and bias of non-response, J. Amer. Stat. Assoc.,48,743-772.
6. Das, A.C. (1950) : Two-dimensional systematic sampling and the associated stratified and random sampling , Sankhya, 10,95-108.
7. El-Bardy, M.A.(1956) : A sampling procedure for mailed questionnaire, J. Amer. Stat. Assoc.,51,209-227.
8. Erikkson, S. (1973a) : Randomised interviews for sensitive questions,Ph.D. thesis, University of Gothemburg.
9. Erikkson, S. (1973b) : A new model for RR, Internat. Statist. Rev., 1,101-113.
10. Folsom, R.E., Greenberg, B.G.,Horvitz, D.G. and Abernathy, J.R.(1973): The two alternate questions RR model for human surveys, J. Amer.,Stat. Assoc.,68,525-530.
11. Hansen, M.H. and Hurwitz, W.N.(1946) : The problem of nonresponse in sample surveys, J. Amer. Stat. Assoc., 41,517-529.
12. Hartley, H.O. and Rao, J.N.K.(1968) : Sampling with unequal probabilities and without replacement, Ann. Math. Stat. 33,350-374.
13. Hartley, H.O. and Ross, A.(1954) : Unbiased ratio type estimators, Nature,174, 270-271.
14. Horvitz, D.G. and Thompson, D.J. (1952) : A generalisation of sampling without replacement from a finite universe, J. Amer. Stat. Assoc., 47. 663-685.
15. `Kish, L. and Hess, I. (1959) : A replacement procedure for reducing the bias of non-response, The American Statistician, 13,4,17-19.
16. Kuk, A.Y.C. (1988) : Estimation of distribution functions and medians under sampling with unequal probabilities, Biometrika, 75,1,97-103.
17. Kunte, S. (1978) : A note on circular systematic sampling design, Sankhya C, 40,72-73.
18. Madow, W.G. (1953) : On the theory of systematic sampling III, Ann. Math. Stat., 24,101-106.

19. Midzuno (1952) : On the sampling design with probability proportional to sum of sizes, Ann. Inst. Stat. Math.,3,99-107.
20. Murthy, M.N. (1957) : Ordered and unordered estimates in sampling without replacement, Sankhya,18;379-390.
21. Murthy, M.N. (1964) : Product methods of estimation, Sankhya,26,A,69-74.
22. Olkin, I. (1958) : Multivariate ratio estimation for finite populations, Biometrika,45,154-165.
23. Politz, A.N. and Simmons, W.R. (1949,1950) : An attempt to get the "not at home" into the sample without callbacks, J.Amer. Stat. Assoc., 44,9-31 and 45,136-137.
24. Quenouille, M.H. (1949) : Problem in plane sampling, Ann. Math. Stat., 20, 355-375.
25. Quenoulle M.H. (1956) : Notes on bias in estimation, Biometrika,43,353-360.
26. Rao, J.N.K., Hartley, H.O. and Cochran, W.G. (1962) : A simple procedure of unequal probability sampling without replacement, Jour. Roy. Stat. Soc., B24, 482-491.
27. Rao J.N.K.,Kovar, J.G. and Mantel, H.J. (1990) : On estimating distribution functions and quantiles from survey data using auxiliary information, Biometrika,77,2,365-375.
28. Royall, R.M. (1970) : On finite population sampling theory under certain linear regression models, Biometrika,57,377,387.
29. Sethi, V.K. (1965) : On optimum pairing of units, Sankhya B, 27,315-320.
30. Singh, D., Jindal, K,K. and Garg, J.N. (1968) : On modeified systematic sampling, Biometrika, 55,541-546.
31. Singh, D. and Singh, P. (1977) : New systematic sampling, Jour. Stat. Plann. Inference, 1, 163-179.
32. Srinath, K.P. (1971) : Multiphase sampling in nonresponse problems, J. Amer. Stat. Assoc., 16, 583-586.
33. Shrivastava ,S.K. (1967) : An estimator using auxiliary information, Calcutta Statist. Assoc. Bull., 16, 121-132.
34. Thompson, S.K. (1990) : Adaptive cluster sampling, J. Amer. Stat. Assoc., 85, 1050-1059.
35. Thompson, S.K. (1991a) : Stratified adaptive cluster sampling, Biometrika, 78, 3089-3097.
36. Thompson, S.K. (1991b) : Adaptive cluster sampling: designs with primary and secondary units, Biometrics, 47, 1103-1105.
37. Warner, S.L. (1965) : Randomised response : A survey technique for eliminating evasive answer bias, J. Amer. Stat. Assoc., 60,63-69.
38. Yates, F. (1948) : Systematic sampling, Phil. Trans. Roy. Soc., London, A 241, 345-371.

Books

1. Chaudhuri, A. and Mukerjee, R. (1988) : Randomised response theory and technique, Marcel Dekker Inc.
2. Cochran, W.G. (1977) : Sampling techniques, Wiley Eastern Limited.

3. Des Raj and Chandok, P. (1998) : Sampling Theory, Narosa Publishing House, New Delhi.
4. Hajek, J. (1981) : Sampling from a finite population, Marcel Dekker Inc.
5. Konijn, H.S. (1973) : Statistical Theory of sample survey design and analysis, North-Holland Publishing Company.
6. Murthy, M.N. (1967) : Sampling Theory and methods, Statistical Publishing Society, Calcutta.
7. Sukhatme, P.V.,Sukhatme,B.V.,Sukhatme,S. and Asok,C. (1984) : Sampling theory of surveys with applications, Iowa State University Press and Indian Society of Agricultural Statistics, New Delhi.

Index